The Passenger Pigeon

THE PASSENGER PIGEON

Errol Fuller

PRINCETON UNIVERSITY PRESS
Princeton and Oxford

Copyright 2015 © by Princeton University Press

Published by Princeton University Press, 41 William Street, Princeton, New Jersey 08540

In the United Kingdom: Princeton University Press, 6 Oxford Street, Woodstock, Oxfordshire OX20 1TW

press.princeton.edu

All Rights Reserved

ISBN 978-0-691-16295-9

Library of Congress Control Number: 2014938023

British Library Cataloging-in-Publication Data is available

The publisher would like to acknowledge the author of this volume for providing the camera-ready copy from which this book was printed.

Printed on acid-free paper ∞

Printed in China

10 9 8 7 6 5 4 3 2 1

(*Half title*). An adult Passenger Pigeon with its young, photographed in the aviaries of Charles Otis Whitman during 1896 or 1898.
(*Title page*). Aquatint by John James Audubon showing a pair of adult Passenger Pigeons.
(*Facing page*). A Passenger Pigeon feather, actual size, drawn by Catherine Wallis (2013). Pencil on paper.
(*Overleaf*). A stuffed Passenger Pigeon.

*Like a star you have lived and like a star you must fall
when the firmament is shake'd.*

> Lord Chief Justice John Popham (1531-1607),
> at the trial of Sir Walter Raleigh

*Tribe follows tribe and nation follows nation like the
waves of the sea. It is the order of nature, and regret is
useless. Your time of decay may be distant, but it will
surely come.*

> The reported words of Suquamish chief Seattle
> (1786 -1866), spoken on March 11, 1854

CONTENTS

PROLOGUE	9
INTRODUCTION	10
THE ANNALS OF EXTINCTION	12
IMAGINE	20
THE BIRD	28
THE DOWNWARD SPIRAL	48
EXTINCTION: THE CAUSES	70
THE LAST CAPTIVES	90
MARTHA	110
ART AND BOOKS	122
QUOTATIONS	148
APPENDIX: A MAGNIFICENT FLYING MACHINE	162
ACKNOWLEDGMENTS	170
FURTHER READING	172
INDEX	175

Prologue

The story of the Passenger Pigeon reads like a work of fiction. At the start of the nineteenth century these birds existed in unimaginable numbers – billions upon billions. The species may have made up as much as 40 percent of the bird population of North America. It may even have been the most numerous bird species on the planet. The flocks were so large and so dense they blackened skies, blotted out the sun. But by the century's end it was over; the birds were gone from the wild. North America's commonest bird had simply vanished. By the year 1914 just a single individual (out of all the countless millions) was left. She was called Martha and she lived alone in a cage at the Cincinnati Zoo. In September this last representative of her species died, and as a living entity the Passenger Pigeon was no more. Along with the Woolly Mammoth, the Dodo, and the Tyrannosaur, the species had become one of the great icons of extinction.

(*Facing page*). A stuffed Passenger Pigeon on a French turned wooden stand.

(*Overleaf*). An immature Passenger Pigeon with other birds, photographed in the aviaries of Charles Otis Whitman, probably by J. G. Hubbard. This is one of a series of photos usually said to have been taken in Chicago during 1896, but evidence suggests some were actually taken in 1898 at Woods Hole, Massachusetts. Courtesy of the Wisconsin Historical Society.

Introduction

This work is not intended as a textbook or a detailed monograph covering every aspect of Passenger Pigeon research and every known piece of information about the species. Anyone requiring this might profitably consult the works by William Mershon (1907), Arlie Schorger (1955), or Joel Greenberg (2014). The present book is simply a celebration (perhaps an inappropriate term in the circumstances), in both words and pictures, of the former existence of the Passenger Pigeon, produced around the time of the 100th anniversary of the death of the last known individual. It is hoped that it will serve as an intriguing memorial to a species that was once so important to the ecology of North America, and will help to bring awareness to just how fragile the natural world can be.

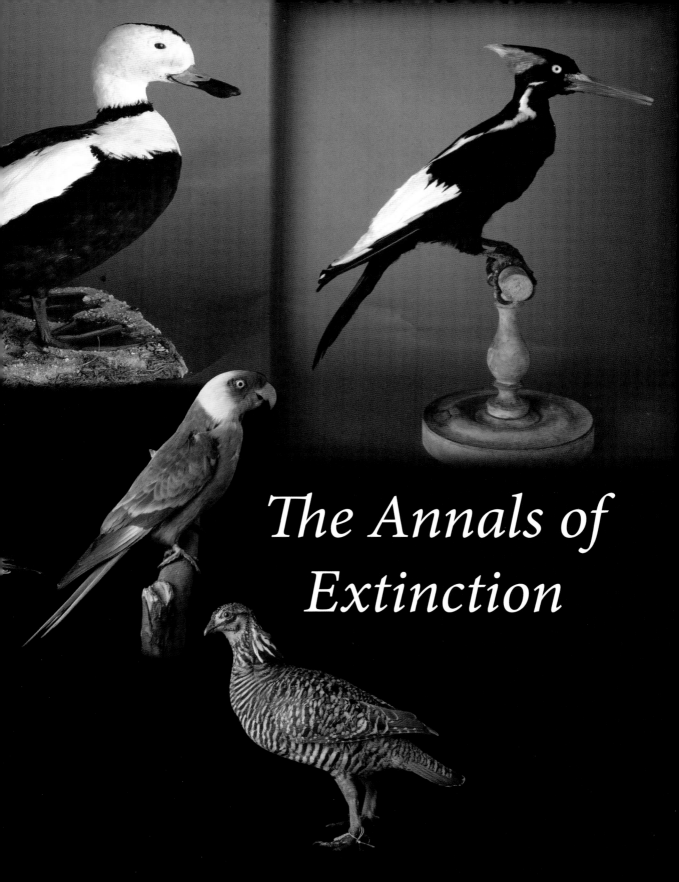

The Annals of Extinction

The Annals of Extinction

All six of the species illustrated on the two previous pages – the Great Auk, the Labrador Duck, the Ivory-billed Woodpecker, the Carolina Parakeet, the Heath Hen, and the Eskimo Curlew – were once inhabitants of North America. In the places they chose to live they were plentiful. All, with the possible exception of the Labrador Duck, were common when Europeans first arrived in the Americas. Two – the Labrador Duck and the Great Auk – became extinct during the nineteenth century; the others managed to live on into the twentieth. Today the status of all six is the same: They exist only as museum specimens. Yet each has an interesting and curious story, and each story could fill the pages of a fascinating book.

The annals of extinction are full of many curious tales, some of them dramatic, most full of tragedy, many mysterious. Why, for instance, did the dinosaurs suddenly vanish after ruling the earth for more than 100 million years? Does the Ivory-billed Woodpecker still cling to existence in remote bottomlands of the southern United States? Probably not, but there are those who believe it does.

(*Previous two pages*). Six species of extinct North American birds.
(*Clockwise from far left*). A Great Auk killed on the island of Eldey, Iceland, during the summer of 1832. A Labrador Duck discovered in 1947 in a nineteenth-century case of stuffed birds at a country house in Kent, England; how it ended up in this locality is unknown. An Ivory-billed Woodpecker. A Carolina Parakeet. A Heath Hen. (The preceding three are museum specimens without significant data.) An Eskimo Curlew, killed at Hadey Harbour, Labrador, during 1862.

(*Facing page*). One of the last Ivory-billed Woodpeckers. This photo was taken by James Tanner on March 6, 1938, and shows Tanner's friend J. J. Kuhn playing with a young individual they named Sonny Boy. Sonny Boy was later released and grew to adulthood; both men saw him again a year or so later. The picture is reproduced through the kindness of Tanner's widow Nancy.

The Annals of Extinction

As for tragedy, what story could be starker than that of the Great Auk, the large black-and-white seabird that in northern oceans took the ecological place of a penguin? Its tale rises and falls like a Greek tragedy, with island populations savagely destroyed by humans until almost all were gone. Then the very last colony found safety on a special island, one protected from the ravages of humankind by vicious and unpredictable ocean currents. These waters presented no problem to perfectly adapted seagoing birds, but they prevented humans from making any kind of safe landing. After enjoying a few years of comparative safety, disaster of a different kind struck the Great Auk. Volcanic activity caused the island

refuge to sink completely beneath the waves, and surviving individuals were forced to find sanctuary elsewhere. The new island home they chose lacked the benefits of the old in one terrible way. Humans could access it with comparative ease, and they did! Within just a few years the last pitiful remnant of this once-plentiful species was entirely eliminated.

(*Facing page*). Digging for a dinosaur at a quarry near Ten Sleep, Wyoming. The bones shown here belong to a *Diplodocus*. Photo by Raimund Albersdoerfer.

(*Above*). *Great Auks in the Mist – A Last Stand* by Errol Fuller (1998). Oil on panel, 33 inches × 44 inches (84 cm × 112 cm). Private collection.

Then there is the most famous of all the extinction stories that have occurred within historical times – the wiping out of the Dodo. The bird's enormous beak and general form may seem a little ridiculous to our eyes, but this was a creature perfectly adapted to its environment, perfectly adapted, that is, for as long as its island kingdom remained undisturbed and unchanged. But, unfortunately for the Dodo, this idyllic state of affairs was not destined to continue. During the year 1598 humans landed for the first time on its remote island in the Indian Ocean, bringing with them powerful predatory creatures: dogs, cats, monkeys, rats. Within three quarters of a century (perhaps considerably less!) the Dodo was gone, and a legend was born.

There are many, many stories like these, all worthy of the telling. But there is one that stands out from the rest, a story so remarkable, so intense, that its elements strain credibility to its limits. It is the story of the Passenger Pigeon, and it is a tale that has everything: great drama, tragedy, intrigue, violence, mystery. Nor is this a story rooted in a dim and distant past where details are sketchy and hard facts have to be supplemented with endless speculation. This is almost a modern tale, and its events are recent enough to have come down to us in great detail. There is even a wealth of photographic material showing living individuals, and these help to give the tale a real sense of immediacy. They almost bring the bird back to life.

Almost, but not quite.

(*Facing page*). *George Edwards' Dodo* by Roelandt Savery, painted circa 1626. Oil on canvas, 32 inches × 40 inches (81 cm × 102 cm). The Natural History Museum, London.

Imagine

Imagine it is some time early in the nineteenth century. We can pick out any year, it really doesn't matter. So let us make it 1810. And let us suppose that you, the reader, have hewn from the wilderness a small area of land. Gradually, you have tamed and cultivated it, and now you are enjoying the fruits of season after season of hard work. You grow enough food, and rear enough livestock, to feed your growing family. There is even a surplus with which you can supply the fast-increasing local community.

The scene could be anywhere in the eastern parts of North America, but let us chose a state, just at random. Let us say that you are somewhere in Pennsylvania. It is an afternoon in May, and things are looking good. Perhaps it is too early to say for certain, but the year's harvest promises to be a splendid one.

You stand in the center of one of your fields recalling with some satisfaction, and not a little pride, the back-breaking effort that you and your family have put in during the bitter winter months and the spring that followed them. As you lean back on your spade you grow conscious of a strange, far-off, almost imperceptible sound, a sound entirely unfamiliar. Unable to decide whether it is a rustle or a buzz, you peer in the direction from which it seems to come. Your gaze passes over the fields to your small orchards, which at last begin to show signs of bearing a decent crop. Then it moves to the forests that surround the farm on all sides, but there is nothing to see; at least there is nothing out of the ordinary. So you turn your attention back to the afternoon's work, but only for a moment. The noise continues, and it begins to distract you from the job at hand. Although still far off, it is surely getting louder, and now it seems more like a drumming than

(*Previous two pages*). Detail from *Passenger Pigeons* by Julian Pender Hume. Acrylic on paper (see page 74 for full image).

a buzzing. Louder and louder it becomes, until all your attempts to ignore it and get back to work come to a complete halt. The sound is certainly coming your way and coming fast. No longer does it sound like drumming; now it is more akin to distant thunder, but with this difference: It is a continuous wall of sound rather than something lasting for just a few seconds.

Suddenly, a few birds, pigeons, appear overhead. Your first thought is that they are fleeing before the ever-increasing racket, and you start to feel some alarm. What catastrophe could cause birds to fly so fast in a frantic attempt to escape? Then you realize that this first thought was wrong. More and more pigeons are passing overhead, and you find it is the pigeons themselves that are responsible for the noise. It becomes truly deafening. As more and more and more of them come pouring in, the numbers are so great the sky itself begins to darken. Within a minute or two it is no longer possible to pick out individual birds; the multitude forms one dark, solid block. The sun is blotted out.

The black mass wheels about. It seems to turn as one unit, not as millions of individual creatures. You have never contemplated numbers of this magnitude before. It is a numerical concept beyond your experience or imagination. And the sound! Your eardrums seem ready to burst. Perhaps the ocean roars like this during a hard storm at sea, but you don't know. You've never been aboard an oceangoing vessel. Now something else happens. The great flock has circled and the pigeons are landing on trees in the forest. Those nearer are coming to rest in your orchards. There seems no end to them. More and more are coming in and landing on the overloaded branches, already packed black with squabbling birds. Droppings fall from the sky like big melting snowflakes. Some are falling on your head! A new sound trumpets across the fields, the sound of splitting timber. The weight of the massed pigeons is so great that here and there it is too much for the trees; their branches can no longer take the strain and they crash to the ground.

(*Overleaf*). *Falling Bough* by Walton Ford (2002). Watercolor, gouache, pencil and ink on paper, 60 inches × 119 inches (154 cm × 303 cm). Reproduced courtesy of the artist.

24

There is nothing to do now but retreat in despair to the shelter of the house. Fortunately, the roof holds little attraction for the pigeons, and largely speaking they avoid it. After a brief period of inaction you venture out, taking your gun with you. After all, a dozen or so cooked pigeons will provide for the family. The gunshots do nothing to scare off any birds, but at least you have a good evening meal.

Three or four days pass. Then, as suddenly as they came, the pigeons are gone. Vanished. Did they return from whence they came, or have they passed on to new pastures? You don't know, and you don't really care. There are far more important things to worry about. The growing crops are destroyed, the buds are eaten or trampled, the orchards wrecked. It is too late in the year to plant again, and the harvest that promised so much will now be a disaster. There will be little to feed the family and nothing to sell to local people. Nor will there be anything left for the livestock. The well is fouled, and this will mean a long walk to the river to fetch fresh water. The damage the birds have wrought can hardly be measured. An entirely new start will be needed – if, that is, you can survive the next few months and the winter that will follow.

The plain truth is that the lives of Passenger Pigeons and technological humankind were incompatible. The species was a great force of nature, with a life cycle in conflict with the new order that was forming in the United States. Its flocks swept across parts of North America like plagues of locusts, seemingly at random, landing *en masse* in places that had food potential. Once the vegetation was stripped the flock moved on to the next chosen place. As the human hold on the North American continent became more extensive, it was inevitable that Passenger Pigeons couldn't survive in the kind of numbers that existed at the start of the nineteenth century. Yet, what no one could have predicted was the utter annihilation of the species, or the fact that this would occur in little more than a century.

Even into the 1850s there seemed to be no great diminution in numbers; at least there was none that anyone noticed. It wasn't until the 1870s that there was any real appreciation that flocks were thinner and their appearances much less frequent. By the 1890s the flocks were gone and just a few isolated individuals lingered on here and there. A year or so into the new century, and the only surviving representatives of the species were a few captive individuals divided between aviaries in Milwaukee, Chicago, and Cincinnati. And the very last one of these died during the first day of September in 1914.

PASSENGER PIGEON
Order—COLUMBÆ   Family—COLUMBIDÆ
Genus—ECTOPISTES   Species—MIGRATORIUS
National Association of Audubon Societies

The

Bird

PASSENGER PIGEON (*Columba Migratoria*)

Upper bird, male ; lower, female

The Bird
(*Ectopistes migratorius*)

So much was written about the Passenger Pigeon during the years when it was a living bird that it is easy to assemble and summarize basic ornithological information about the species. Yet despite the wealth of written evidence, there are still questions about certain aspects. Did the birds regularly lay one egg or two? Did they nest just once in a given year, or did they reproduce multiple times? Did their vast numbers protect them in some way against the predatory assaults of other species? Different accounts provide varying answers for these and other mysteries. Perhaps essentially unresolvable questions are natural with even the most familiar of extinct creatures. Perhaps it is fortunate that as far as the Passenger Pigeon is concerned so many of the main facts are firmly in place, and a fairly comprehensive account can be given.

The name that science has bestowed on the species is *Ectopistes migratorius* – simply a description of its lifestyle expressed in Latin. Roughly translated it means "migratory wanderer." The common name is derived from the French word *passager*, which means "to pass fleetingly."

These were birds that could be recognized at a glance as pigeons, yet even so they were somewhat aberrant in form. Worldwide there are more than 300 species in the pigeon and dove family (Columbidae), and

(*Previous two pages*). Three images of the Passenger Pigeon: adult male, chick, and an adult pair.
(*Left*). The adult male. An illustration by E. J. Sawyer taken from an educational leaflet published by the National Association of Audubon Societies (circa 1920).
(*Center*). A photograph of a chick. Before the advent of color photography it was common practice to tint black-and-white images with color. This image has been reversed from its original state (reproduced on page 45).
(*Right*). The adult pair (male above, female below). This watercolor was painted by Louis Agassiz Fuertes for the frontispiece of William Mershon's celebrated monograph on the species, published in 1907. Its present whereabouts are unknown.

Description

Length: 15 - 16 inches (39 - 41 cm)

Male: Head bluish gray; area at back of neck iridescent bronze, green, or purple depending on angle of light; back slate gray tinged with olive brown; lower back and rump grayish blue becoming grayish brown on uppertail coverts; two central tail feathers brownish gray, rest white; wing coverts brownish gray with irregular subterminal spots; primaries and secondaries darker grayish brown, secondaries edged with white; throat and breast rich pinkish rufous, becoming paler on lower breast to white on abdomen and undertail coverts; bill black; iris red; naked orbital ring purplish red; legs and feet red.

Female: Similar to male but duller overall, with reduced iridescence; head, neck, and back grayish brown; underparts buff brown and less rufous; tail shorter; legs and feet paler red; iris orange red; naked orbital ring grayish blue.

Immature: Similar to adult female but with scapulars, wing coverts, and feathers of foreneck and breast tipped with white, and with no iridescence; legs and feet dull red; iris brownish surrounded by narrow ring of carmine.

Measurements
Male: wing 196 - 215 mm; tail 175 - 210 mm; bill 15 - 18 mm; tarsus 26 - 28 mm.

Female: wing 180 - 210 mm; tail 150 - 200 mm; bill 15 - 18 mm; tarsus 25 - 28 mm.

most of them look quite similar, though there are some exceptions. In appearance the most outlandish and untypical was the Dodo (*Raphus cucullatus*), which, perhaps rather surprisingly, was actually a gigantic flightless member of the family. At the other end of the scale, *Ectopistes migratorius* was quite possibly the most elegant and stylish of all pigeons.

Judging by appearance alone, the Passenger Pigeon seems to be closely related to the still common but much smaller Mourning Dove (*Zenaida macroura*), but recent DNA analysis suggests that its nearest relationship is not with this species. Apparently, its kinship lies instead with pigeons of the genus *Patagioenas*. In North America this genus is represented by the Band-tailed Pigeon (*P. fasciata*).

(*Above*). Two close relatives of the Passenger Pigeon.
(*Left*). The Mourning Dove (*Zenaida macroura*). Although much smaller, this species is remarkably similar in appearance to the Passenger Pigeon, yet DNA analysis suggests it is not its closest relative.
(*Right*). Pigeons of the genus *Patagioenas* are now thought to be more closely related, although their general appearance makes this difficult to believe. The genus is represented in North America by the Band-tailed Pigeon (*P. fasciata*), pictured.

Basically a streamlined version of the normal rather dumpy pigeon shape, the Passenger Pigeon had very long wings, a long, graduated tail, and a slender body and small head. Clearly, it was designed for speed and endurance when on the wing. This natural elegance of form was enhanced by beautiful yet subtle feather coloring. Even the gleaming iridescence on parts of the plumage was rather understated, yet it is easy to visualize how glorious this opalescence must have looked glinting in the sun, the effect multiplied a billion times.

These birds were nomads, forever on the move in search of the vast quantities of food needed to support their uncountable hordes. Seeds and fruits made up most of their diet, but they also gobbled up worms, insects, snails, and other small invertebrates. In particular they liked hard mast,

(*Above*). Passenger Pigeons, immature (*left*), adult male (*center*), adult female (*right*).Watercolor painted by Louis Agassiz Fuertes for E. H. Eaton's *Birds of New York* (1910 - 1914).

forest nuts that occur in great quantities in certain places during some years, but prove more plentiful in other places at other times. They were very fond of beechnuts and chestnuts and, amazingly, could swallow acorns whole.

The birds' mouths and throats had tremendous elasticity, and a special joint at the corners of the lower bill allowed for an enormous increase in capacity. In relation to the size of the bird, a great quantity of foodstuff could be stored in the crop, causing the neck to bulge in an extraordinary manner before food was fully swallowed and digested. Were an individual to be shot with a cropful of nuts it would hit the ground with a rattle that one writer likened to the sound of a bag of marbles being shaken.

Curiously, the chestnut trees that supplied a fair amount of the mast on which Passenger Pigeons thrived became virtually extinct themselves due to a fungus imported from Asia around the year 1905. The trees were unable to withstand this new enemy, and it has been estimated that as many as 30 billion died during the course of a few decades. This rapid decline was too late to affect the pigeons, however; by 1905 they were already gone from the wild.

Many descriptions of the arrival of the great flocks and their behavior when feeding have been penned. One that particularly intrigued Christopher Cokinos when he was conducting research for his book *Hope Is the Thing with Feathers* (2000) came from the painter Don Eckelberry (1921 - 2001) during an interview. Toward the end of his life Don was often referred to as "the grand old man of American bird painting," and in many respects he provided a real and fascinating link with the past. He was, for instance, probably the last person to make a genuine sighting of a living Ivory-billed Woodpecker (*Campephilus principalis*), and he loved nothing more than to regale visitors with his memories of old-timers and their relationships with birds. During the interview with Mr. Cokinos, Don recalled talking many years previously to a ninety-year-old farmer near Cleveland. This farmer described how he'd seen what he called "big blue pigeons" coming in great bunches and landing at the forest edge where the beech mast was. They would keep moving ahead, with the back of the flock constantly flying up and over the rest, the flock forever pushing

A distribution map for the Passenger Pigeon compiled by Arlie Schorger in the early 1950s. The solid line outlines the area of normal distribution, the dotted line encloses the typical nesting areas, and the solid dots show casual or accidental distribution.

forward in a great rolling motion. "Sort of rolling ahead," was the old man's description.

Since mast is such a variable commodity (plentiful in one particular place in one year but perhaps scarce the next), the species' reliance on it as a staple diet poses a question. How did the birds know where to go to find it? The answer is, of course, that we don't really know. Chief Simon Pokagon of the Potawatomi had an interesting way of putting this in words:

[They] have communicated to them by some means unknown to us, a knowledge of distant places.

Passenger Pigeons certainly had keen eyesight, and their great powers of flight enabled them to survey vast areas in their relentless searching. This, surely, was what the wandering was all about. They were constantly seeking places that could offer enough food for a short or long stay, together with the possibility of shelter for an equal length of time.

The species' range was essentially linked to the eastern deciduous forests of North America, from the Great Plains east to the Atlantic, north to southern parts of Canada and south to northern Mississippi. Some birds went farther afield: In winter flocks could be found in the extreme southeast of the United States, and some were seen even in Mexico and Cuba. Passenger Pigeons occasionally went way off course, and there are records of individuals being spotted in Europe. Such records are difficult to substantiate, however, and most are probably bogus or occurred due to the release and subsequent sighting of captive individuals.

The vast aerial migrations of Passenger Pigeons were connected with two different activities: the need to find roosting sites and the need to find places to nest. There is a great deal of difference between the two. The term *nesting* is pretty much self-explanatory; birds choose a site that affords them the opportunity and facilities to build nests and raise young. It implies a stay in the chosen area of some time; in the case of the Passenger Pigeon this period would be for perhaps four, five, or even six weeks.

Roosting sites were simply places where birds could shelter overnight. The stay was, therefore, for an indeterminable period of time. It may have been for just a night or two, or sometimes for much longer. Mostly this would depend on the amount of food available relative to the number of birds in a given flock. It might also be dependent on the severity of

(*Facing page*). For museum collections, birds are often preserved as flattish reference specimens (often known as "cabinet skins") rather than as fully stuffed birds positioned in a lifelike manner. This has the advantage of making the specimen easily available for scientific study (measuring etc.), as well as facilitating easy storage. This particular example is still in perfect color, having been stored in darkened quarters and not exposed to sunlight, which causes fading. There is an allegation that it was shot in England, but this provenance may be false or may have resulted from the shooting of an escaped captive bird. Photo by Roddy Paine Photography.

persecution by humans, weather prospects, or variables of which we have no knowledge. Naturally, roosts would differ in size. Some were comparatively small, and the area the birds used might be no more that a few acres in extent, whereas others could be truly vast, covering 100 square miles or more. From such sites the birds would forage in the surrounding countryside. Some reports suggest that they were capable of making round trips of 200 miles (sometimes even longer) during the course of a single day, leaving a roost early and returning at nightfall or even later. There are instances in which a great flock of birds would use the same roosting area year after year. In other situations they would come once and never return.

Nesting was a very different matter. Because of all the tales of fantastically vast nesting colonies, it is perhaps surprising to learn that these birds sometimes nested in single pairs or small groups. However, the vast majority of stories that come down to us do indeed tell of colonies containing millions of individuals. A very late record of a colony that descended on an area near Sparta, Wisconsin, in 1871 tells us that it spread across some 850 square miles. There are even details concerning its general form: It was L-shaped, with a few areas left untouched. Why the birds didn't favor these particular areas is not made clear.

In southerly latitudes the swarm arrived early, sometimes coming in March; farther north it was a little later. The nest itself was built quickly, some reports suggesting that it was made and the eggs were laid within three days. Compared to the nests of many bird species, the Passenger Pigeon's nest was a crude affair, usually made from around eighty or ninety twigs loosely assembled into a flimsy, straggling shape just strong enough to support a single egg, sometimes two. (There is some doubt and argument over what was the norm; if all the records are compared, it seems virtually certain that the usual number was one.) With so many individuals collecting twigs at the same time, it was sometimes said that the ground beneath the nests looked as though it had been swept clean, but this observation seems hard to accept when the vast quantities of dung being constantly dropped is taken into account.

There has been some debate among ornithologists over the kind of trees the species favored, and also the sort of terrain they liked best:

maples rather than oaks or beech, wet areas as opposed to drier ones, sandy stretches rather than boggy places. The truth seems to be that these birds would take advantage of any place that offered the facilities to support

(*Above*). Nest and egg of a Passenger Pigeon thought to have been photographed in the aviaries of Charles Whitman during 1896 or 1898.

(*Overleaf*). Two views of an abnormally pale-plumaged Passenger Pigeon alongside a typical specimen. The mutation that causes this coloration is comparatively common in wild birds, yet it is interesting that this is the only known example among more than 1,500 Passenger Pigeon specimens that exist in the world's museums. It was once part of the collection of Walter Rothschild, a man who was particularly interested in color aberrations, and at his death in 1937 it was bequeathed to London's Natural History Museum. Today it is housed in that institution's ornithology department at Tring, Hertfordshire, England. The photograph comes from "Colour Aberrations in Some Extinct and Endangered Birds," a paper soon to be published in the *Bulletin of the British Ornithologists' Club,* and is used by kind permission of the authors, Julian P. Hume and Hein van Grouw.

their vast numbers. In other words, they may very well have had favorites, but they weren't restricted to one particular type of habitat.

Just as there are differences of opinion over this, so too there is controversy over how many times during a year the bird bred. The best evidence seems to suggest only once, but there are a number of claims that it happened twice or more.

The eggs took thirteen days to hatch, and both parents helped with incubation. Reports suggest that the male took a turn from mid-morning to mid-afternoon and the female put in the rest of the time. Once the chick hatched, both parents fed it with "pigeon milk," a curd-like substance made in the adult bird's crop.

(*Facing page*). A captive at its nest in the aviary of Charles Whitman, probably photographed by J. G. Hubbard, in 1896 or 1898.

(*Above, top*). A Passenger Pigeon egg allegedly collected in Utah during 1881. If the locality data is correct, this would have been a most unusual occurrence. The egg is shown at more than twice the natural size to indicate just how plain it was.

(*Above, bottom*). A Passenger Pigeon egg taken in 1865, shown at natural size.

After about six days adult foods began to be gradually introduced. When about two weeks had passed, a most remarkable thing happened. The adults gave their chick one last feed, then rose *en masse* from the nesting site and flew off like a great moving cloud. Quite simply, the chicks were abandoned. They stayed in the nest for a day or two, and when all hope of the adults returning seemed gone, they climbed from the nest and dropped to the ground. The very fat little creatures began to move forward, avoiding or crossing any obstacle in their path. Obviously at this time they were extremely vulnerable, and many would fall victim to predatory mammals and birds. Several days later they could fly well enough to escape from any persecutors, and away they went.

One of the more extraordinary things that can be found in Passenger Pigeon literature is a series of musical notations recording the sounds that the birds made in different circumstances. These notations were made by a gentleman named Wallace Craig and published in the *Auk* (vol. 28, no. 4, 1911), the journal of the American Ornithologists' Union. Craig wrote them down while observing some of the last living individuals at the aviaries of Charles Whitman in Chicago during 1903. In his summary of his efforts, he said:

> *The following notes are put forth, not with the assurance that they adequately represent the repertory of this remarkable species, but only with regret that the meagre information now to be given is all we are likely to have on the subject.*

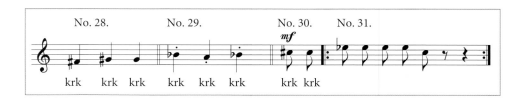

(*Above*). Some female calls, reproduced in musical notation by Wallace Craig during 1903.

(*Facing page*). A Passenger Pigeon chick. This photo is one of a series taken during 1896 or 1898 in the aviaries of Charles Whitman, and seems to have been taken during the month of August.

The Bird

45

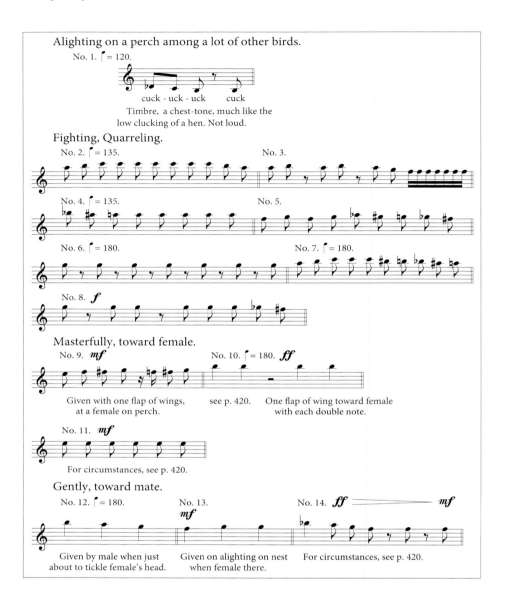

(*Above and facing page*). More of Wallace Craig's musical interpretations of Passenger Pigeon vocalizations.

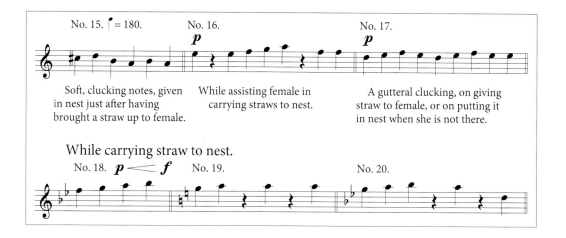

There are various other written accounts of the calls the birds made. Most tell of fairly ugly shrieks and croaks, but some describe a twittering and cooing that, because of the vast numbers, could be heard for miles. John Mactaggart (1829) gave a rather romantic description:

> *The singular noise they make in their crowded nursery…surpasses any sound I have ever heard. It is a loud and confused buzz of love.*

Another writer was less flattering and said the sound was like the croaking of wood frogs. Perhaps the most poetic description mentioned that the pigeons had no real song but instead gave out:

> *A number of low notes, some of which are sounds that seem to be almost the soft breathing of the great trees.*

The Dowward Spiral

National Association of Audubon Societies for the Protection of Wild Birds and Animals

THE PASSENGER PIGEON

This bird is now believed to be extinct. Many other valuable species are threatened with a like fate. We are trying to preserve them.

The Downward Spiral

Everything leads to the belief that the pigeons, which cannot endure isolation and are forced to flee or to change their way of living according to the rate at which North America is populated by the European inflow, will simply end by disappearing from this continent, and, if the world does not end this before a century, I will wager...that the amateur of ornithology will find no more wild pigeons, except those in the Museums of Natural History.

Bénédict Henry Révoil (1856)

The Passenger Pigeon needs no protection. Wonderfully prolific, having the vast forests of the North as its breeding grounds, travelling hundreds of miles in search of food, it is here today, and elsewhere tomorrow, and no ordinary destruction can lessen them.

Report of the Ohio Senate (1857)

The first European to make any written comment on the Passenger Pigeon was a French explorer by the name of Jacques Cartier (1491 - 1557), who made a note of them on July 1, 1534. Cartier is widely credited with having claimed Canada for the French and, perhaps because he had weightier things on his mind, his written account is brief and reveals little. It was soon followed by the observations of other writers, however, some of whom were well-known historical figures, Samuel de Champlain (1567 - 1635) and Cotton Mather (1663 - 1728) among them.

(*Previous page*). The cover of a leaflet issued by the National Association of Audubon Societies promoting the protection of wild birds and featuring the Passenger Pigeon as an icon to attract attention to its work (circa 1920).

Most of the descriptions given by early explorers and colonists tell of the wonder that the vast numbers of these birds inspired, and the idea that one day they might no longer exist was entirely foreign to such men. The whole concept of extinction was not one with which they were familiar; curiously, it is a relatively modern thought. Just as the dominant seventh chord in music (now so easy for us to listen to) would have sounded discordant to their ears, the idea that a creature might die out completely was a leap of the imagination too great for them to take. Extinction was a concept not yet formulated; it was an idea for the future. If a species ceased to exist in one particular area, there was a general assumption that it could still be found somewhere else. The notion that it was gone from the world, never to return, was completely alien. Even if it had occurred to these early pioneers, in the case of the Passenger Pigeon the evidence of their senses would have suggested otherwise, and they would have felt compelled to believe that the vast numbers could never diminish.

The awe and wonder that flights of Passenger Pigeons inspired in the early observers is briefly summed up in the words of a certain Thomas Dudley (1574 - 1653), deputy governor of the Massachusetts Bay Colony. On March 28, 1631, he penned a letter from Salem to his old mistress back in England. She was the Lady Bridget, Countess of Lincoln:

> *Upon the eighth of March…there flew over all the towns in our plantations, so many flocks of doves…that they obscured the light…It passes credit, if but the truth be written.*

And then, in the gloriously flamboyant language of Jacobean and Carolean Britain, he finished with an ominous, even fearful, flourish:

> *But what it portends, I know not.*

Rather more explicit and detailed is a passage originally written in French by Pierre Boucher (1664) just a few years later. It can be translated as follows:

> *There are such prodigious quantities of them [pigeons] that forty or forty five can be killed with the single discharge of a gun. This is not done ordinarily, but it is common to kill eight, ten, or twelve…The Iroquois take them with nets as they fly by, sometimes catching three to four hundred at a draw.*

Through the seventeenth and eighteenth centuries there seemed little alteration in the health and status of the species. Descriptions tend to tell largely the same kind of story – countless numbers of birds flying overhead, skies darkened, great hauls of dead pigeons (shot, knocked down, netted, or otherwise destroyed), dung and destruction everywhere, flocks coming from nowhere and vanishing as mysteriously as they came. A record from May 1771 tells of a single day at a Boston market when some 50,000 dead pigeons were being offered for sale.

Even given such horrendous statistics, there is no written hint that there could be any overall change in the status quo, although looking back with the full benefit of hindsight, it becomes apparent that the species was being pushed back (albeit slowly) westward.

During 1731 Mark Catesby (1683 - 1749) produced the first published illustration of a Passenger Pigeon for his monumental and celebrated work *The Natural History of Carolina, Florida, and the Bahama Islands* (1731 - 1743). All of Catesby's illustrations are rather crude and quaint, and his Passenger Pigeon is no exception, but there is no mistaking the distinctive shape and form. Catesby showed little regard for relative proportion, and in this case he perched his bird on truly gigantic leaves of the turkey oak.

The nineteenth century dawned and the pigeon population seemed as vast as ever. The era of the great amateur naturalist was beginning, and men such as Alexander Wilson (1766 - 1813) and John James Audubon (1785 - 1851) experienced at first hand the flights of these birds. Wilson arrived in America a penniless immigrant from Paisley in Scotland but, after failing in attempts to be taken seriously as a poet, he is known today as the "father of American ornithology." For his celebrated book *American Ornithology* (1808 - 1814) he made a serious attempt to estimate the number involved in a great flight of birds that he observed. Wilson wrote:

> *From right to left as far as the eye could reach, the breadth of this vast procession extended…Curious to determine how long this appearance would continue, I took out my watch to*

(*Above*). The watercolor painting from which Mark Catesby developed the illustration for his famous book *The Natural History of Carolina, Florida, and the Bahama Islands*. All of Catesby's original watercolors were acquired by the British royal family, which still owns them.

The Passenger Pigeon

note the time, and sat down. It was then half past one. I sat for more than an hour, but instead of a diminution of this prodigious procession, it seemed rather to increase both in numbers and rapidity...I rose and went on. At almost four...in the afternoon I crossed the Kentucky River...at which time the living torrent above my head seemed as numerous... as ever...all moving in the same...direction, till after six in the evening...If we suppose this column to have been one mile in breadth (and I believe it to have been much more), and that it moved at the rate of one mile in a minute; four hours, the time

(*Above, left*). John James Audubon, painted by John Syme in 1826. Oil on canvas, 35 inches × 27 inches (90 cm × 69 cm). The White House Collection, Washington.

(*Above, right*). Alexander Wilson, a black-and-white image after an oil painting attributed to T. Sully.

> *it continued passing, would make its whole length two hundred and forty miles…Supposing that each square yard comprehended three Pigeons, the square yards in the whole space, multiplied by three, would give two thousand, two hundred and thirty millions, two hundred and seventy two thousand pigeons (in figures this translates as 2,230,272,000)! An almost inconceivable multitude, and yet probably far below the actual amount.*

When John James Audubon published his famous book *The Birds of America* (1827 - 1838), it came without text. This was due to a complexity of British publishing law requiring an author to surrender to British institutions several copies of any book containing text. As this would have meant a serious financial burden, Audubon took the simple precaution of omitting any. To make up for this lack he wrote *Ornithological Biography* (1831 - 1839), which was much less costly to produce, and published it entirely separately. Clearly, he had to write about Passenger Pigeons, and not willing to be outdone by Wilson's eccentric but charming calculation, he made his own:

> *Let us take a column one mile in breadth, which is far below the average size, and suppose it is passing over us without interruption for three hours at the rate of one mile in the minute. This will give us a parallegram of 180 miles by 1, covering 180 square miles. Allowing two pigeons to the square yard, we have one billion, one hundred and fifteen millions, one hundred and thirty six thousand pigeons in one flock.*

The absolute precision of these quaint calculations must obviously be taken with a pinch of salt, but their accuracy – or lack of it – does not really matter. They are simply reflections of the great wonder the tremendous flights inspired, and various other writers came up with similar numerical formulas. All of them remain to give us some idea of the enormity of the population, and several modern writers have been equally unable to resist the temptation to engage in their own arithmetical exercises. For his monumental monograph *The Passenger Pigeon*, published in 1955, Arlie Schorger based his sums on aremarkable description of the birds given by Major W. Ross King (1866). In this description King mentioned a flight that lasted for fourteen hours in one day and continued in diminished form for several days afterward. Schorger arrived at a calculation suggesting 3,717,120,000 birds were involved.

More recently, Christopher Cokinos tried to give the matter a different perspective for his fascinating book on the extinct birds of North America, *Hope Is the Thing with Feathers* (2000). Basing his arithmetic on Wilson's numerical estimate, he calculated that if the individual birds had flown beak to tail in single file they would have stretched around the earth's equatorial circumference 22.6 times!

It might be borne in mind that each of these calculations was based on the flight of a single Passenger Pigeon colony. How many such colonies were in existence is something we just don't know. No truly meaningful comparative studies were ever made. Perhaps there were very few. Perhaps there were more than this. Any attempt to arrive at a meaningful answer would be speculation. But whatever the truth, as the middle of the nineteenth century approached there seemed to be no perceptible decrease in the species' numbers. And this was despite the wholesale slaughter, the cutting back of the forests, and the fast-increasing human population.

Monsieur Bénédict Révoil appears to have been a lone voice when he expressed his fear that the Passenger Pigeon was doomed (see the head of this chapter), and even his opinion was not caused by any appreciable diminution in numbers. It was simply based on a commonsense analysis of the situation following his witnessing of a massacre that took place in Hartford, Kentucky, in the autumn of 1847. Clearly, the species was being steadily pegged back westward (with several observers noting that it was much less frequent, or absent altogether, in the far east), but whether or not there was any great decline in overall numbers is a moot point.

It wasn't until the late 1850s that a decline began to be commented on, and even then any fall in numbers seemed hardly worthy of note. There were still millions and millions of birds. A decade later, however, it was

The Downward Spiral

(*Above*). *Passenger Pigeons in Flight* by Lewis Cross (1937). Oil on canvas, 4 feet × 11 feet (122 cm × 307 cm). Lakeshore Museum Center, Muskegon, Michigan.

(*Overleaf*). Spot the Passenger Pigeon. A typical ornamental display case from the first half of the nineteenth century revealing the perceived lack of significance then attached to Passenger Pigeon specimens. Surrounded by several other common North American species, a Passenger Pigeon perches inconspicuously at the right of the display. These preserved birds were allegedly taken to Britain by John James Audubon when he was trying to drum up support for the publication of *The Birds of America* and, after the inclusion of a European Wood Pigeon (center), given as a gift to a British well-wisher. Photo by Roddy Paine Photography.

becoming increasingly apparent that something was afoot, and from 1870 onward the slump in numbers became truly precipitous. Within just a few more years the vast population had declined to virtually zero.

This terrible downward rush to oblivion presents an unanswered, perhaps unanswerable, question. Was the slide really sudden (that is, had the population previously been able to constantly replenish itself despite the dreadful levels of persecution), or had the unimaginable numbers been dropping dramatically, though to human eyes largely imperceptibly, for decades? The historical record certainly shows that the species was pressed back from more easterly haunts, and that this was noticeable even during early colonial times. Presumably, the very gradual and erratic westward progression was a continuous one, but any significant awareness of its implications became submerged beneath the wealth of descriptive prose detailing the comings and goings of the vast multitudes.

When people wrote about the Passenger Pigeon, as so many of them did, they concentrated their efforts on the spectacular rather than making any meaningful attempts to trace geographic decline. And despite the many wonderful written accounts, this was always a species steeped in a curious kind of mystery to those who saw it. Where did it come from? Where did it go to? Would the horde come again? What damage had it done? How could advantage be taken of the great flights? These were the questions that occupied the minds of those who encountered the species.

As far as the actual decline is concerned, the logical conclusion must surely be that Passenger Pigeon numbers had been reducing for many years, but the sheer vastness of these numbers tended to fool the eye and the mind. Eventually there came a point when the slump toward zero began to become apparent, even though there were still millions of individuals left.

And when the tipping point came, the fall was unbelievably fast.

Following a nesting near Shelby, Michigan, during 1874 (one of the very last of the great nestings), an observer using the pseudonym of Tom Tramp (1876) wrote the following:

> *With the number of pigeons left, the steadily increasing demand for them dead or alive, and the fearful increase in*

the rank and file of professional netters who follow them by telegraph hundreds of miles, it is high time to do something for their protection... There were probably as many, or more, pigeons caught this year than ever, but it was not on account of the increased number of birds; it was done by the more numerous army than ever of pigeoners.

Tom Tramp estimated that 25,000 dead pigeons were being shipped out daily for a period of four weeks.

(*Above*). *Pile of Passenger Pigeons* by Maria E. Mangano. Ink and watercolor on paper, each bird measuring approximately 0.5 inches × .75 inches (1.25 cm × 2 cm). Courtesy of the artist, who retains full copyright for any use of this image.

Horrific accounts come down to us concerning the aftermath of the massacre at Shelby. In addition to the general slaughter, thousands of birds were taken alive and kept in great pens, presumably in the hope that their meat would remain fresh and available once any glut in the market had subsided. The hope was forlorn on two counts. First, the birds were housed in such appalling conditions that most quickly perished from a lack of food and water, or else just fretted themselves to death. Second, the actual glut was just too great. Chicago markets were so overstocked that the price of a whole barrel of pigeons dropped to less than fifty cents. Even then, many went unsold, and their unused bodies simply rotted away.

The nesting at Shelby represented a defining event, for the dramatic fall was suddenly almost complete. But there were still a few nestings to come, and two of them were truly spectacular. The first of these occurred in Warren County, Pennsylvania; the arrivals began on March 7, 1878, and the birds continued coming in for the next two weeks or so. The regular tale of plunder followed, and estimates suggest that over a million birds were shipped to markets in New York and elsewhere.

But this great nesting was dwarfed by one that occurred around Petoskey, Michigan, and started just a few days after the Warren County one. The Petoskey nesting has gone down in history as the last mighty flourish of the species, and its size was staggering; accounts indicate it was spread across an area more than 100,000 acres in extent. It is a curious fact that by this time the state of Michigan (and also Wisconsin) had enacted laws giving some protection to Passenger Pigeons. However, these laws were not clearly framed and proved difficult to enforce. Judged from any practical standpoint, they were utterly ineffective.

A few stalwart wildlife protectionists led by a musician named H. B. Roney made efforts to stop the inevitable slaughter, but their well-meant

(*Facing page*). A cabinet skin prepared from a bird killed at the great nesting at Petoskey. The data label reads: "Bought of Mr. Elkington, Petoskey Michigan, 1878."

attempts met with little real success. Roney's estimate of the number of birds that died may be exaggerated, but it gives some idea of the scale of the butchery. His guess was that a billion birds were killed, either directly or indirectly. In January 1879 he wrote that when he and his assistants arrived at the scene:

> [We] *stood and gazed in bewilderment…[as the birds] darted hither and thither…in every direction, crossing and re-crossing…before the dizzy eye of the beholder.*

Roney also reported the way in which pigeoners recklessly chopped down trees to get chicks. In order to collect evidence of law violation and try to inhibit the killing, he resorted to various ruses. He spread rumors that the Petoskey birds had eaten poisonous berries and were, therefore, totally unfit for human consumption. He would pretend to fall asleep, snoring theatrically when in the company of pigeoners who were suspicious of his intentions. Roney hoped this would encourage them to speak more freely among themselves and reveal their tactics and the scale of their operations to his ever-open ears. He spoke regularly to children as they seemed more likely to reveal information and innocently implicate their elders in lawbreaking acitvities. He published accounts of pigeoners standing up to their necks in blood and gore, and old women completely denying any knowledge of pigeons while simultaneously frying dead birds killed just a few hundred yards away. His efforts and those of some colleagues brought about a few prosecutions, but it was essentially just a few poverty-stricken trappers who suffered while the larger operators escaped unscathed. There was, of course, a major reaction to Roney's campaign, and he was accused of distorting evidence, spreading lies, exaggeration, and a lack of regard for the poor people who relied on pigeon hunting for sustenance. Interestingly, the Michigan laws were tightened up in 1897, leaving no room for ambiguity. Shooting pigeons at any time was absolutely banned. In discussing this development, Christopher Cokinos (2000) made a rather philosophical point. This was in effect the perfect law, he wrote, because everyone was obliged to obey it: By 1897 there were no pigeons left to shoot!

(*Above*). The cover of a circular issued in early 1879 (and largely reproduced from an article first published in the *Chicago Field*) that was intended as a vigorous retort to H. B. Roney's allegations.

The great nesting in Petoskey was by no means the very last, but later ones, although still large, were pale reflections of what had gone before. During the last two years of the 1870s and on into the 1880s nestings were reported from areas as far apart as Ontario, Michigan, Oklahoma, Texas, Missouri, Wisconsin, and Pennsylvania. Each was characterized by an increased general wariness among the flocks, but there was a continuation of the slaughter nevertheless. It was noticed that it was now quite common for birds to abandon their nestings at the first sign of persecution, but thousands did not.

During the first years of the 1890s there were undoubtedly small flocks of pigeons still in existence, but, curiously, we don't know of this from written accounts of such groups. We know because various records from markets tell of substantial numbers of birds still being offered for sale. The last of these kinds of records seems to come from late 1892 or early 1893 when a quantity of dead pigeons was offered for sale in St. Louis. From here on the tales are essentially of very small groups or of single individuals. Some of these are clearly records of lonely, misplaced birds. A written account made by E. B. Clark (1901) concerns the sighting of a single individual in a Chicago park during 1894. The man who first spotted the bird was appalled when someone approached expressing a wish for a gun. "He had no soul above pigeon pie," observed the gentleman who penned the story for posterity.

Some indication of just how misplaced surviving individuals were is made clear by sampling a few of the last records. Several birds were seen in Franklin County, Iowa; others were spotted in Connecticut and a few more in Tennessee. Some were seen in Indiana, and a lone individual was caught in New Jersey. Another was shot in North Carolina and yet another killed in Melrose, Massachusetts. Early in 1895 two Passenger Pigeons were shot in Louisiana and another met its end in Nebraska. A gentleman named Oliver Jones was responsible for a significant, if rather tragic, milestone in the species' history. On June 21, 1895, near Minneapolis, he collected a male bird, a nest, and an egg, constituting the last reliable human sighting of a nesting, albeit a failed one. The specimens were given to the Bell Museum of Natural History, University of Minnesota (the museum is likely to move in the near future).

A last record for Louisiana has an equally sad ring to it. E. A. McIlhenny, the son of the man who concocted Tabasco sauce, wrote in 1943 of a bird killed in November 1896. He encountered a flock of Mourning Doves and noticed that among them was a lone Passenger Pigeon – which he promptly shot. The suggestion may seem too anthropomorphic for some tastes, but perhaps this sad creature was just desperate for companionship.

Many claims of sightings from the last years of the nineteenth century and the very first years of the twentieth are generally assumed to be false, most of them probably resulting from confusion with the similar-looking Mourning Dove. However, a few are certainly genuine, being supported by specimens that were killed and their stuffed remains presented to institutions. But even these verifiable records of wild birds eventually grind to a halt, and then comes controversy over which of them actually constituted the very last.

For years this rather dubious honor fell to a bird killed by a fourteen-year-old lad named Press Clay Southworth (1885 - 1979) on March 24, 1900, near Sargents, Pike County, Ohio. Press had shot the bird because he did not recognize it; but his parents did. When he showed them the corpse, his mother immediately knew what it was, having seen many of the birds in years gone by. She insisted that the remains be taken to a local amateur taxidermist, Mrs. Charley Barnes. In taxidermical terms Mrs. Barnes operated only at a rudimentary level, and to compound her somewhat indifferent work, she found she had run out of the cleverly made glass eyes that professional taxidermists use to imitate real ones. So she improvised and used buttons.

A few years later, Press and his family presented his historic specimen to the Ohio Historical Society, and it is now kept at the Ohio History Center's Museum in Columbus. The donation was made in October 1915, just a month or so after the death of Martha, the celebrated captive Passenger Pigeon at the Cincinnati Zoo.

Although various garbled versions of Press Southworth's story were published over the years, it is entirely due to the detective work of Chris Cokinos (2000) that accurate accounts can now be given. By the time of Mr. Cokinos' interest in the story, Press himself was dead (although he

had lived well into his nineties), but his granddaughter was located and she supplied the full tale.

Although most books and articles on the subject list Press's quarry as the last fully authenticated instance of a Passenger Pigeon in the wild, Buttons (as she came to be known) was certainly not the last wild bird. For one thing, she seems to have been a young individual, perhaps no more than a year old. Her parents may well have still been in existence at the time she was shot. In any case, Joel Greenberg (2014) has recently brought to light two later records, one of which is equally convincing in terms of authenticity to the Buttons record. A male – one of two individuals that were, apparently, seen – was shot on April 3, 1902, near Laurel, Indiana. Although, like Buttons, the bird was stuffed, it was unfortunately destroyed long ago.

(*Above, left*). The boy who shot the pigeon: Press Clay Southworth photographed at about the age of seventeen, within a few years of his killing of one of the last wild individuals. Courtesy of Mary, Terry, and Ted Kruse, and Christopher Cokinos.

(*Above, right*). Buttons, the pigeon killed by Southworth and one of the last wild birds.

There are a number of later allegations of birds being seen, but none can be properly substantiated. Some, however, are fairly convincing. Among these is a note made by President Theodore Roosevelt (1858 - 1919), no slouch as a naturalist, who believed he saw a small flock in Virginia during May 1907. His report is one among several.

And then they were gone. Apart, that is, from a few individuals still surviving in captivity.

Extinction: The Causes

Silent Wings

A Memorial to the Passenger Pigeon

PUBLISHED BY

The Wisconsin Society for Ornithology

Extinction: The Causes

There need be no doubt that when Europeans began to colonize North America the invasion spelled ultimate doom for the Passenger Pigeon. Final destruction took many, many years, and for much of this time such an outcome may have seemed inconceivable, but it was inevitable. The collision between technological man and Passenger Pigeon was one the pigeons could neither avoid nor benefit from. Their own habits left them entirely vulnerable to the depredations of men with the will and capability to profit from their slaughter and from the destruction of the habitats the birds needed in order to survive.

There are three main elements to the dismal tale. The first is the sheer ferocity of the human onslaught. The second is the wholesale destruction of the forests that pigeons relied on for food and shelter. The third is a lifestyle that left the species utterly defenseless before the twin requirements of the continent's new arrivals: food and land. The bird's very abundance and need to live and support itself in vast colonies brought about disaster.

We can only make guesses, albeit reasonably informed ones, about the relationship Passenger Pigeons established with Native Americans in times long past. The lifestyles of the indigenous peoples were, of course, at variance with those of European colonists, and their populations did not have the advantage of rapid numerical increases caused by a constant influx of newly arriving immigrants.

Obviously, the birds were preyed upon by various tribes long before the arrival of Europeans; such a handy food resource could hardly be ignored.

(*Previous two pages*). The cover of *Silent Wings*, a booklet published in 1947 by the Wisconsin Society for Ornithology as a memorial to the Passenger Pigeon. On it is depicted a scratchboard drawing by Hjalmar A. Skuldt of a monument erected at Wyalusing State Park, Wisconsin, accompanied by drawings by Jacob Bates Abbott of pigeons being shot and netted. This particular copy was given to the well-known Danish ornithologist Finn Salomonsen in 1960 by Paul Hahn, a man who conducted a census of all the Passenger Pigeon specimens in the world's museums and published the results of his findings in *Where is that Vanished Bird?* (1963).

There is no doubt that American Indians regularly adopted the practice of netting pigeons, often in very large numbers. Apparently the Seneca called the Passenger Pigeon *jahgowa*, meaning "big bread," a term that requires little explanation, and there are several accounts of members of the tribe raiding pigeon nestings with great ferocity. But the means of raiding were by no means as sophisticated or as all-pervading as the attacks by European colonists. Presumably in bygone times a situation had long been reached that twenty-first century man might describe as a "sustainable balance."

From the seventeenth century onward the depredations of the colonists were of a different order from those of the indigenous peoples, and they increased in tempo and severity as the years passed. By the mid-nineteenth century they had reached a crescendo. All kinds of stratagems were employed to trap or kill birds, and it is not always easy to tell which was the most destructive.

Shooting is the one most often recounted, and there are many written descriptions of the actual carnage (such was the intensity of the destruction that on at least one occasion a cannon was used). One account can probably stand for all. An anonymous journalist working for a daily Wisconsin newspaper called the *Fond du Lac Commonwealth* wrote an article that was published in the May 20, 1871, issue. In it the writer gave a graphic description of a typical scene:

> *Imagine a thousand threshing machines running under full headway, accompanied by as many steamboats groaning off steam, with an equal number of…trains passing through covered bridges…and you possibly have a faint conception of the terrific roar following the monstrous black cloud of pigeons as they passed in rapid flight…a few feet before our faces…nearly on a level with the muzzles of our guns…the contents of a score of double barrels was poured into their dense midst. Hundreds, yes thousands dropped into the open fields below. Not infrequently a hunter would discharge his piece and load and fire the third and fourth time into the same flock. The slaughter was terrible beyond any description.*

Our guns became so hot by the rapid discharges, we were afraid to load them. Then, while waiting for them to cool…we used…pistols, while others threw clubs…the scene was truly pitiable…many [birds] were only wounded, a wing broken or something of the kind. These were quickly caught and their necks broken.

(*Above*). A painting produced by Julian Pender Hume for the cover of his book *Extinct Birds* (2012). Although different in some aspects of composition, it is deliberately based on on the well-known image opposite. Acrylic on paper, 12 × 18 inches (31 × 45 cm). Private collection.

When shooting birds flying overhead it was often unnecessary even to aim. Dozens of individuals could be brought down with a single shot from a gun simply pointed skyward. A single ammunition supplier recorded handling three tons of powder and sixteen tons of shot during a nesting near Sparta, Wisconsin, in 1871, a statistic that gives a certain dimension to the slaughter. And this dealer would have been just one among many.

(*Above*). An often-reproduced image, taken from *The Illustrated Sporting and Dramatic News* (July 3, 1875) and titled *Winter Sports in Northern Louisiana: Shooting Wild Pigeons*. It was drawn by an otherwise virtually unknown illustrator called Smith Bennett.

Sometimes the securing of the carcasses was very much secondary to the act and the excitement of shooting the living birds. When shootings occurred at nestings this would often result in vast numbers of birds being driven away before the act of raising the young was complete.

Not content with just shooting the creatures in huge numbers, humans often felt the need to formalize and quantify the action. Competitions were organized, and, where such things could be successfully arranged, the activity passed for sport. Sometimes this took the form of releasing captured individuals (often already incapacitated by wounds they had received) from traps and gunning them down as they flew off in the hope of freedom.

While researching the subject of trap-shooting for his 1955 monograph on the species, Arlie Schorger found a little poem in the first volume of

(*Above*). A number of engravings similar to the one on the previous page were published in magazines and journals. This one was in the edition of *Leslie's Illustrated Newspaper* issued on September 21, 1867.

Extinction: The Causes

an obscure Philadelphia publication (*Cabinet of Natural History and American Rural Sports*) of 1831. It was titled *Pigeon Shooting*:

> *Of all the themes that writers ever chose*
> *To try their wits upon in verse or prose,*
> *A pigeon-shooting match would surely be*
> *The last selected for sweet poesy.*

(*Above*). An advertisement for a pigeon shoot from the year 1833.

At other times, trap-shooting was rejected and the competition was reduced to a simple matter of men and women (but usually men) with guns standing at regularly spaced intervals bringing down as many birds as they could while the flocks passed overhead. Then each competitor would make a head count of the birds in his or her own particular pile.

One sad little account details the concern that pigeons seemed to have for wounded comrades. It was published long after the species had passed into extinction, and whether or not it is reliable cannot be said. In 1933 a certain William Manlove wrote:

> *It was pathetic to see the efforts of the companions of a wounded pigeon to support him in his flight. One after another would dart under the stricken one as he began to sink, as if to buoy him with their wings. They would continue these efforts long after he had sunk below the general line of flight, and not until all hope was lost would they reluctantly leave him and rejoin the flock.*

Extinction: The Causes

By no means content with just shooting birds, people employed various other methods of demolition. Netting was one of the most effective, and with a species as populous as the Passenger Pigeon (and one prone to flying in such close formation) it brought fantastic returns. Flying birds were lured to carefully placed nets by a variety of methods. Often they were attracted by foodstuffs placed strategically on the ground. Alternatively, captive individuals flying high on the end of a rope were used to entice wild flocks, and so too were the decoys known as stool pigeons. These poor captives had their eyelids sewn shut to make them effectively blind, and were then placed on a perch just a few feet from the ground. The netters would cunningly control them (and the perch) by means of a cord and wait for a passing flight of birds. Then they would pull on the cord, causing the perch

(*Facing page*). An advertisement from an 1880 edition of *Forest and Stream*.

(*Above*). A pigeon net painted during 1829 in Canada by James Pattison Cockburn.

79

to topple and the stool pigeon to flutter helplessly to the ground. If the ploy worked the pigeon flock would believe the stool pigeon was one of their own making a landing, and follow suit. The trap was sprung.

What followed is easy to imagine. The awful scene would be characterized by desperate, fluttering, and squealing birds (often numbering in the hundreds) making vain attempts to squeeze through the netting that enclosed them. Trappers would simply walk to their nets and snap bird necks and skulls one by one. Some of the more hardened would bite the birds' heads to crush their skulls. To modern sensibilities the sight of all this blood and hideous distress would be repulsive and horrifying.

Another way of achieving some degree of success was knocking down birds from their nests with poles or clubs or using these kinds of implements to bring them down as they flew low overhead. This method was relatively ineffective when compared with simply chopping down trees in order to get at the plump nestlings. There is a record of the speedy cutting down of a huge stand of trees covering 1,500 acres, and this sort of procedure was going on regularly. The number of individuals, both adults and young, destroyed in such an enterprise is perhaps unimaginable, but additional subtlety was even introduced to this process. Pots of burning sulfur were sometimes positioned beneath trees, and the fumes would stupefy birds until they fell down. If flocks nested or settled in birch trees, the papery flammable bark was ignited, and, as the heat rose, the squabs descended.

If long dry grass or other ignitable vegetation occurred beneath nesting trees, humans would set fire to it; many of the pigeons would drop down, overcome by smoke, and their roasted remains could be collected the next day, or once things had cooled.

(*Facing page*). Stereoscopic views were a nineteenth-century invention that featured duplicated images printed side by side. They were printed on a card designed to be placed in a special device (a stereoscopic viewer) that gave the user a convincing 3-D image. These views were highly popular and were usually issued in sets. All kinds of different subject matter were available, and information about the image was usually printed on the card's reverse. This card, showing stuffed Passenger Pigeons, comes from a set promoting natural history.

HURST'S STEREOSCOPIC STUDIES OF NATURAL HISTORY,
For Schools and Parlor Entertainments. No. 4.

CLASS II, ORDER II, PASSERES. Family Columbidæ.
The Wild Pigeon, (Fig. I, female ; II and III, males).

Ectopistes Migratoria, DE KAY, N. Y. Fauna, p. 196.
Columbia Migratoria, LINNÆUS, Syst. Nat., p. 285.

DESCRIPTION.

Male greyish lead color on back part of neck, head and wing coverts, upper wing coverts of a brownish hue, with a few black spots, wings dull brown, edged with rufous. Tail consists of 12 feathers ; two center feathers blackish brown ; on each of the others is an oval black spot close to the base on inner side of shaft, two outside feathers nearly white. Throat, breast and sides of a brownish red, on the sides of the neck a brilliant metallic reflection, bill black, iris bright red, feet pink, claws black, length from 16 to 18 inches.

Female smaller than the male, her colors are less bright and more tinged with brown.

The nest—a layer of a few sticks is made in a single day, the young are hatched in 16 days ; both male and female assisting in making the nest, the former bringing the materials and the latter arranging them. They lay two eggs, pure white ; each brood generally consists of male and female. Wilson observes "that each nest contains but one squab." From my own experience in visiting there nesting places I have oftener found two than one ; Audubon also mentions two squabs to a nest.

Wilson says : " In passing through a deserted breeding place every tree for miles was spotted with nests. In many instances I counted upwards of 90 nests on a single tree." He gives a rough calculation of a flock that he once saw at Frankfort, Ky., which numbered " 2,230,272,000 pigeons, and which extended 240 miles, allowing each pigeon to consume one half pint of food daily, the whole quantity would equal 17,424,000 bushels daily."

The specimens figured are a pair obtained by me in the spring of 1849 or 50 at Utica, N. Y., during a severe snow storm which occurred there in the early part of May. They were then migrating north to Booneville, where they generally built their nests. The snow was so thick and heavy that they were completely bewildered, flocks of thousands were detached from the main body, flying in circuits or shooting around and between the houses. Hundreds lost their lives by "butting" against the Central R. R. Depot, and the gable end of "Bagg's" Hotel. This wholesale suicide not only caused great sport, but feasted the inhabitants on pigeon pot-pie for many a day after.

PHOTOGRAPHED BY HAINES, 478 BROADWAY, ALBANY, N. Y.   ORDERS RECEIVED BY JAMES A. HURST, 9 ELM STREET, ALBANY, N. Y.
Price 50 Cents.

The plain truth is that people would catch and kill pigeons in vast numbers using any method they could devise or find efficacious, and it was the very lifestyle of the species that made it vulnerable to the full measure of human rapacity.

If any additional illumination on the scale of the slaughter is required, a few statistics will provide it. Christopher Cokinos (2000) unearthed some records. A single pigeon dealer in Monroe County, Wisconsin, shipped two million birds to market during 1883. In another instance the entire floor of a large warehouse was covered in pigeon corpses to a depth of three or four feet. A calculation was made suggesting that three Michigan nestings during 1875 yielded 1,000 tons of squabs. Such figures do not even begin to take into account the thousands of birds that, for a variety of reasons, were left on the ground.

Meat was obviously the main motivation for all this killing, but there were subsidiary reasons. A buttery substance was made from oil and fat obtained from the birds. Fifty pigeons would, apparently, yield a pound of feather bedding. Perhaps there are even pillows stuffed with Passenger Pigeon feathers that survive and are still in use – who knows?

Maria Rundell, in a book called *American Domestic Cookery* (1819), gave a recipe that proposed a very delicate way of preparing and serving the birds for the table:

> *Make [two pigeons] look as well as possible by singeing, washing, and cleaning the heads well. Leave the heads and the feet on, but the nails must be clipped close to the claws. Roast them of a very nice brown, and when done put a little sprig of myrtle into the bill of each... The head should be kept up, as if alive, by tying the neck with some thread, and the legs bent as if the pigeon sat upon them.*

It is doubtful that many birds were treated to this prissy yet grisly fate. Most would have been cooked and eaten in much more prosaic fashion. Indeed, another cookbook suggested that good flavor could be achieved only if dead birds were cropped and drawn soon after they were killed. It added that no other bird needed so much washing!

Although it is the cause most regularly cited, direct hunting was by no means the only pressure brought to bear on pigeons. Nor was it necessarily the most critical one.

The destruction of huge areas of forest made survival for creatures that needed to live in vast colonies impossible, and the scale of forest clearance in areas frequented by Passenger Pigeons is staggering. Woodland and forest were removed primarily to make land available for agriculture or townships, but even places that weren't required for such purposes were subject to habitat destruction as their trees were chopped down for fuel or to provide building materials. The demolition had been going on at a furious rate since Europeans began to colonize the continent, and its intensity increased as the years rolled by. Between 1850 and 1910 around 180 million acres were cleared for farming alone, a figure that equates to more than thirteen square miles every day for a period of sixty years. Large areas of woodland currently cover eastern North America, and it might be pointed out that wildlife of all sorts can flourish there. But nowhere are there the kind of forests needed to support billions of Passenger Pigeons. Unlike its relative the still-common Mourning Dove, it is evident that the species had developed a way of living entirely dependent on existence in vast numbers. No other numerical equation would work.

There are probably a whole range of very practical reasons for this need, many of which we are unable to define. Some can be identified, however. The strange way in which adults abandoned their young is one of these. When the chicks were around two weeks old the adults just flew off, leaving their offspring alone in the nest. After a while the abandoned creatures simply flopped to the ground, and it was a day or two before they were themselves able to fly off. It is easy to imagine what happened during this brief period. The youngsters were distressingly vulnerable, and mammalian, reptilian, and avian predators would come from miles around to take advantage of these defenseless babies. With the vast number of squabs available, however, this local predatory population soon became sated, and in percentage terms only a small proportion of pigeon chicks were actually consumed. If, on the other hand, there were only a few birds at a nesting and, therefore, a comparatively small number of

The Passenger Pigeon

Extinction: The Causes

squabs, it seems highly likely that local populations of predators could quickly mop up the whole lot, once the young had dropped to the ground.

Whatever the full set of reasons for the need to live in great colonies (and whatever vulnerabilities and protections this afforded), the fact remains that it was the twin blitzkrieg of hunting and deforestation that brought about annihilation.

Curiously, there are those who have been unable, or unwilling, to accept the obvious and have found it necessary to come up with whole rafts of eccentric explanations. Among these exotic solutions is the suggestion of mass drownings. The idea was proposed that billions of birds were driven off course by bad weather and ended their lives over the Atlantic or the Gulf of Mexico. Alternatively, it was suggested that they could have been lost over the Great Lakes. There is no doubt this sort of thing did indeed happen, but it had no bearing on the extinction of the species. A curse delivered by an aggrieved Christian minister was among the more extreme proposals put forward. And some people remained in complete denial over the whole issue. Passenger Pigeons had simply gone elsewhere, they thought. Chile, for reasons unknown, was considered to be a likely haven. Someone even decided the pigeons brought about their own downfall by laying white eggs! Epidemic avian disease (*Trichomoniasis*, pigeon canker, for instance) was a more plausible explanation, but not one for which there is any evidence.

In reality the extinction of the Passenger Pigeon is not quite the mystery it appears to be, and the causes can be identified by using common sense. Reasons for the actual decline are obvious: overhunting and habitat destruction. But in terms of absolute extinction, an extra factor comes into play: the peculiar evolution of the birds that meant they could survive –

(Previous two pages). A habitat display at the Denver Museum of Nature and Science featuring Passenger Pigeons. Although this splendid and extravagant display shows how a group of birds might have appeared in natural surroundings, the truth is that the species could never have survived in such small groups.

(Facing page). Two paintings by the New Zealand born artist Raymond Ching, depicting rare species that, unlike the Passenger Pigeon, have been able to survive in tiny populations for many years. (*Left*). A Takahe. Oil on panel. (*Right*). A Kakapo, a noctural parrot that also happens to be one of the world's rarest bird species. Oil on canvas.

psychologically, emotionally, and physically – only in vast numbers. There are many examples of bird species in which the requirement for large colonies is not necessary, and some instances of exceptionally rare animals managing to cling to existence for decade after decade in amazingly small numbers. The extraordinary New Zealand Kakapo (*Strigops habroptilus*), for instance, is the world's largest parrot. It is also the only one that is nocturnal and flightless, and it has managed to survive as a species in tiny populations for well over a century. By the time of the death of Martha, the Kakapo had already seemed doomed to imminent extinction for decades past. Yet it somehow survived through the whole of the twentieth century and then on into the twenty-first, albeit in pitifully small numbers. A similar tale could be told of another rare and celebrated New Zealand bird, the Takahe (*Notornis mantelli*), rediscovered after having been feared extinct for many years. But these were species with different requirements than the Passenger Pigeon, and a quite different destiny.

Once Passenger Pigeon numbers had dropped below a certain level (no matter how high that level may have been) the species was doomed. Whether the number was 50 million, 10 million, 500,000, or even lower, we shall never know, but it is certain there was an actual tipping point and that at some moment in the second half of the nineteenth century that tipping point was reached.

In historical times North America was heavily forested east of the Great Plains, and this entire area was the playground of the species. It moved through the rich environment taking advantage of all the bounty it offered. Winters were spent touring southern forests, and as summer came there was a northward movement to the Great Lakes, southern Canada, New England, and New York, and all the time the great stands of forest were an essential requirement. As long as this environment remained stable, survival was secure, and the birds could rely on warm summers and plentiful rainfall to produce abundant supplies of food from more than a million square miles of woodland. But North America has always been a work in progress, and the Passenger Pigeon evolved and then declined according to that progress.

Times change, and a species requiring half a continent to roam over freely could not hope to carry on in radically altered circumstances. Although for a while the interests of humans were served by the presence of Passenger Pigeons, eventually the survival of the birds proved incompatible with their needs.

The arrival of technological man, the ultimate super-predator, on North American shores heralded the beginning of a new order. It was an event that a species operating in the manner of the Passenger Pigeon had no hope of withstanding.

(*Facing page*). An engraving published in a general book on natural history: *The Royal Natural History*, by Andrew Wilson (circa 1880).

The Last Captives

The Last Captives

The tale of the captive birds is mostly a grim one. The majority of captured individuals were taken (and then kept alive) for specific purposes, and those purposes were usually horrific. They were to be used either as stool pigeons or other kinds of decoys, or else they were kept alive until the day (usually one not far off) when they could be killed to provide fresh meat.

A few individuals were more fortunate and took a place in aviaries and zoos, but as the species was so common for so long, these captives aroused little interest. Few records were kept. The London Zoo, for instance, once published a list of all the species it had ever kept up to the year 1927 (Carmichael Low, 1929); the fact that it once owned individuals is clearly listed, but no details are given, nor do any photographs seem to have been taken. John James Audubon claimed he took several hundred living birds to Britain and deposited most of these at the London Zoo, but this claim seems extraordinary.

During the 1890s the interest in captive members of the species increased. The big problem was that by now, of course, there were very few, and by the end of the century the surviving captive individuals were divided among three collections, with one small group in Milwaukee, another in Chicago (moved each year to Woods Hole, Massachusetts, and then back again), and the third in Cincinnati.

The least known of the three is the small colony kept in Milwaukee in a fairly rudimentary structure made of wood, glass, and wire. This aviary

(*Previous two pages*). Passenger Pigeons with birds perhaps belonging to other species, probably photographed by J. G. Hubbard in the aviaries of Charles Otis Whitman. This is one of a series of photos usually said to have been taken in Chicago during 1896, but evidence suggests that some might actually have been taken in 1898 at Woods Hole, Massachusetts. Courtesy of the Wisconsin Historical Society.

(*Facing page*). A photograph from the same series showing an adult male. Courtesy of the Wisconsin Historical Society.

The Passenger Pigeon

was made by a man named David Whittaker who seems to have maintained it for no other reason than his own general interest. Beginning the collection during 1888, he acquired two pairs from an American Indian who had caught them near Shawano Lake, Wisconsin. Soon after the acquisition, one bird scalped itself by flying at speed into the wire of its cage, and another somehow escaped, leaving just a pair. This remaining pair managed to breed, and within a few years Whittaker had a flock of fifteen individuals. Some aviculturists enjoy great success with their captives, while others do not; Mr. Whittaker seems to have been one of those who had what would be described in gardening terms as a green thumb. He realized at an early stage that feeding his birds with a supplement of insects and worms would increase their health.

Ruthven Deane (1851 - 1934), a prime figure in ornithological research at the time, found out about this small colony and, realizing its potential importance, made a detailed study of it, the results of which he began to publish in 1896. It is natural to assume that after lives spent in captivity these birds would have developed some degree of tameness, but this seems not to have been the case. Given his undoubted skill at rearing captive birds, it seems possible that Whittaker had deliberately pursued a policy of leaving them relatively undisturbed, and as a result they never became used to the presence of humans. Whatever the cause, Deane found them wary and skittish in the extreme. He gives an interesting description of their behavior as rough weather approached Milwaukee:

> *Old birds...arrange themselves side by side on the perch, draw the head and neck down into the feathers and sit motionless for a time, then gradually assume an upright position, spread the tail, stretch each wing...and then, at a given signal...spring from the perch and bring up against the wire netting with their feet as though anxious to fly before the disturbing elements.*

(*Facing page*). Another of the photographs of one of Charles Whitman's birds, this one showing what is described on the photo's reverse as "a very perfect specimen of the adult female." Courtesy of the Wisconsin Historical Society.

During 1896 a university lecturer named Charles Otis Whitman (1842 - 1919) began turning his attention to Passenger Pigeons. Whitman's university career was quite untypical, and very varied, for a man of his time. For a while he worked in Leipzig, then in Tokyo, and subsequently in Naples before returning home to the United States. One of his particular interests was the study of pigeons, and this fascination led him to begin purchasing birds from Whittaker's Milwaukee aviary and adding them to a collection of other species at his own premises in Chicago.

On March 14, 1896, he bought three birds, a male and two females. Later, in October, he bought an additional pair. Then in March of the

(*Above*). Charles Otis Whitman photographed about 1901 with some of his captive pigeons. None of those in the picture seem to be Passenger Pigeons. Photographer unknown.

following year he acquired the balance of the flock; from these, nine young were produced, of which four survived. At its maximum his flock numbered nineteen individuals, and it was probably in the light of this initial success that during 1898 he gave seven individuals back to David Whittaker. Presumably this was done in the vain hope that a division of the livestock might make it more vigorous. At about this time Whitman began to take his birds to Woods Hole, Massachusetts, where he had become the first director of the Marine Biological Laboratory.

Apparently, the long trip to Massachusetts by railcar became a yearly event for the birds. They were taken from Chicago in early summer and returned in the autumn, a disturbance that was surely not conducive to successful breeding.

The twelve birds Whitman had left soon became fourteen, but from this point onward any numerical increase was marginal, During February, March, and the first part of April 1899 (around a year after he'd given the seven birds back to Whittaker) his remaining captives laid no fewer than fifteen eggs – but none of them hatched. Four more eggs were laid during the last days of April, but although three of these did hatch, none of the chicks survived.

More eggs were laid during 1900, but again no young were raised. Then there was a short period of hope, because by 1902 the flock had risen to sixteen in number. During that year two things happened, however. A female (an individual that may have been the famous Martha) was sent to Cincinnati, and two elderly females escaped from the Woods Hole pen. Of the thirteen individuals now left, eight were males and five were females. From here on it was a sorry story of steep decline, and by 1907 there were just two surviving birds. Both were female, and both died during the winter. Curiously, Whitman had kept his Passenger Pigeons with other species (as can be seen clearly in some of the existing photos), and all that now remained were two male hybrids, crosses between a Passenger Pigeon and a female Ring Dove (*Streptopelia risoria*). What became of these two is not known. According to the writer Wallace Craig (1913), who spent time at the Chicago aviaries, Whitman penned a detailed account of his birds, but this seems never to have been published, nor is its location known.

The Passenger Pigeon

(*Above*). A rear view of a Passenger Pigeon with an Australian species, the Crested Bonzewing (*Ocyphaps lophotes*), as a companion. One of the photos shot in Charles Whitman's aviaries, it shows beyond any shadow of a doubt that different pigeon species were kept together in the same cage. In the photo, the Bronzewing is in rather better focus than the Passenger Pigeon. Courtesy of the Wisconsin Historical Society.

It was while the birds were seemingly flourishing in Whitman's aviaries that a series of fascinating photographs was taken. There are twenty-four in all (although more were undoubtedly taken), and their survival into the present time is remarkable in itself. A few of them were reproduced and published in journals, magazines, and books during the first decade of the twentieth century, but most went unused. Fortunately, copies were somehow acquired by Frank M. Chapman (1864 - 1945), a man who worked for many years at the American Museum of Natural History in New York, and eventually these were passed on to the Wisconsin Historical Society, an institution that has carefully preserved them and made them available for study.

There is a certain amount of doubt over who actually shot the photographs and exactly where and when they were taken. Several differing proposals have been suggested. On the back of the surviving photos someone (probably Frank Chapman) has scribbled the year 1896, and at least one of them has 1896 inscribed onto the image. However, an often overlooked article published in volume 15 of *Bird-Lore* (March –April, 1913) contradicts this evidence. In this issue of the journal (then the official organ of the Audubon Societies) a few of the photos were reproduced, and the editor acknowledged, with gratitude, the permission he'd been given to use them. He wrote:

> *They were made by Mr. J. G. Hubbard, who generously contributed them to* Bird-Lore, *at Woods Hole, Massachusetts, in the summer of 1898.*

So it seems that some were taken in Chicago during 1896, but the majority were probably shot at Woods Hole two years later. Also, one or two may actually have been taken by Charles Whitman himself.

During the summer of 1907, just as Whitman's birds were dying out, Ruthven Deane went back to Milwaukee to make a further study of the birds David Whittaker had kept, and to see how the returned individuals were faring. What he found was depressing. Only four, all of them males, remained alive; the last surviving female had died during the previous year. Care of the captives had now passed to a Mr. A. E. Wiedring, but there seems

The Last Captives

to be no record of why or when this change had happened. Perhaps advancing years were limiting David Whittaker's ability to look after his charges, but if it was hoped that the change would improve the lot of the birds, this hope had proved to be forlorn.

(*Facing page*). This photo of two of Charles Whitman's birds is captioned with the words "parent and young." Clearly inscribed on the image is the number 96. Presumably this refers to the year in which the picture was taken. Courtesy of the Wisconsin Historical Society.

(*Above, left*). Passenger Pigeon with a companion. Courtesy of the Wisconsin Historical Society.

(*Above, right*). A solitary individual. Courtesy of the Wisconsin Historical Society.

(*Previous four pages*). Photographs taken in the aviaries of Charles Otis Whitman showing Passenger Pigeons in a variety of poses, sometimes alone, sometimes with companions, sometimes with birds of other species. It is known that Whitman allowed his captives to hybridize, and some of the images may show hybrid individuals. All photographs are courtesy of the Wisconsin Historical Society.

(*Above*). Passenger Pigeons quartered with birds of other species. Courtesy of the Wisconsin Historical Society.

The Last Captives

(*Above*). A Passenger Pigeon perched on a wooden platform and dappled in streaks of sunlight. Scribbled on the back of this photograph is the comment that the image shows the species in a highly characteristic pose. Courtesy of the Wisconsin Historical Society.

The Passenger Pigeon

Photo by C. O. Whitman (University of Chicago)

October 16, 1906.

Mr. W. B. Mershon,

Dear Sir:—I am much chagrined over my carelessness in overlooking your request for a photo of a young Passenger Pigeon. I had best of intentions, but crowded work threw this out of mind. I should have attended to it at first, had it been easy to get at the picture. I have been away all summer and found things misplaced on my return. I fear it is now too late, but send the picture to be used if you are still able to do so. I shall be very much interested to see your book. I still have two female pigeons and two hybrids between a former male pigeon and the common Ring-dove. The hybrids are unfortunately infertile males. Very truly,

C. O. Whitman.

The Last Captives

Between November 1908 and February 1909 these remaining individuals, all bred from David Whittaker's original pair, died allegedly from tuberculosis, although old age seems just as likely a cause.

There are a number of poorly documented stories of individuals still surviving in the hands of backwoodsmen, but if these are discounted as the tall tales they probably are, the only captive individuals were now the remnants of a group that had been living in Cincinnati.

The Cincinnati Zoo and Botanical Garden opened to the public during September 1875, making it the second oldest zoo in the United States. It seems that right from the outset it housed Passenger Pigeons. Quite why it had them at this time is not known. After all, the species was, then, still perceived as common, and as an exhibit the birds must have aroused little interest when compared with Cincinnati's other residents, which included wombats, bears, alligators, a talking crow, and an elephant. Whatever the reason for the species' inclusion, there may have been a flock of more than twenty captives. A few more were added during the next few years; apparently they were all kept in a cage measuring ten feet by twelve. Although these were creatures that enjoyed being close to others of their kind, these confined quarters must have been distressing for birds designed so specifically for flight.

It seems that after the initial success in getting new birds, attempts were occasionally made to acquire more, but by and large these proved impractical. The individual acquired from Charles Whitman during 1902 (the female that may actually have been Martha) appears to have been the only late addition to the flock.

Gradually the Cincinnati birds died, and by 1909 there were just three left. And one of them was Martha.

(*Facing page*). Facsimile of a page from William B. Mershon's monograph *The Passenger Pigeon*, published in 1907. Mershon reproduced a note from Charles Whitman beneath the reproduced image, and the remarks speak for themselves; obviously, Mershon had time to squeeze the photo into his book. This image does not seem to be among those obtained by the Wisconsin Historical Society and indicates that there were rather more pictures than the ones presently known. Whether or not this particular photo was taken by Whitman himself is something of a moot point.

Martha

Martha

One death is a tragedy, a million deaths are a statistic.

Josef Stalin (1879 - 1953)

There is no doubt that from all the billions of Passenger Pigeons that ever lived, only one has achieved any kind of individual identity. And she is Martha. The rest form just a great horde, a massive homogeneous block that once existed, but exist no longer.

Yet even Martha is an enigma. All that is really known about her is that she lived and died, that her death occurred in a cage at the Cincinnati Zoo, that her stuffed remains are now at the Smithsonian Institution in Washington, and that she was probably the very last of her kind. Those who write about her sometimes mention that her death represents the only instance when the exact moment of a species' extinction is on record, but this notion is wrong on a number of counts, just one of which is enough to scotch it entirely. The truth is that we don't actually know when Martha died, at least not with any degree of exactitude. This lack of definitive information is largely due to differing accounts given by the main keeper at the zoo, Salvator "Sol" Stephans, and his son Joseph. Both kept changing their story.

One of their accounts states that Martha died in her cage at precisely one o'clock on the afternoon of Tuesday, September 1, 1914. Another suggests her death might actually have occurred some four hours later. One version of the story relays the romantic idea that she died surrounded by

(*Previous two pages*). Martha, as she is today, in close-up. Detail from the photograph reproduced in full on page 119, showing the intricate beauty of her plumage. Photograph by Donald E. Hurlbert. Courtesy of the Smithsonian Institution, Department of Vertebrate Zoology, Bird Division, National Museum of Natural History.

(*Facing page*). Martha in her cage at the Cincinnati Zoo. The identity of the photographer is uncertain and is thought to be either Enno Meyer or William C. Herman.

a group of grieving keepers. Another maintains that just Sol and Joseph were with her. Yet another is less poetic and indicates she died alone and was found lying on the floor of her cage by an assistant keeper named William Bruntz. The truth hardly matters, of course. Dead is dead.

Furthermore, any serious attempt to go through written zoo records to solve the mystery is impossible; they were destroyed by fire in 1963. What we do know is that the first day of September was a swelteringly hot day with high humidity levels; perhaps it was all just too much for a frail old bird.

So what else is there? The answer is, not a lot. It seems certain that Martha was born in captivity, but where? Mr. Stephans' recollections were as varied on this subject as they were on others. Perhaps this was due to poor note keeping; perhaps it was forgetfulness or just a lack of knowledge concerning past events that at the time of their occurrence would have seemed unimportant. One suggestion he made was that she had been hatched at the zoo and had lived there for upward of twenty- five years, but on another occasion he stated categorically that this wasn't the case and that she was originally one of the flock kept by Charles Otis Whitman in Chicago. As it is on record that Whitman did give a female bird to the zoo in 1902, it is more than possible that this very individual was Martha.

However all this may be, there seems little reason to doubt that the last Passenger Pigeon did go by the name of Martha, but the facts behind how she acquired the name are as debatable as many of the other details of her life. Her last surviving cage mate was called George, and it is usually recorded that the pair were named after George Washington and Martha, his wife. Yet even this little tale is by no means certain. In one of his stories, zookeeper Stephans relayed the information that Martha was named after the wife of one of his friends. Whatever the truth of this, it is highly likely that George and Martha became the last living pair of Passenger Pigeons during 1909 (probably in April) when another Cincinnati captive, an old male bird, died. By this time all of Charles Whitman's Chicago flock was gone, and so too were the remnants of David Whittaker's Milwaukee birds. Nor are there any reliable records of any individuals still living in the wild.

In 1909 Martha and George were alone. There are a few descriptions of their time together. Apparently, they often called to each other, making sounds that visitors described as "see, see." Sometimes they made loose, untidy nests in a tree that stood inside their aviary.

(*Above*). Martha, the last Passenger Pigeon, in life. This photograph has been reproduced many times in books, magazines, and journals. It was taken by Enno Meyer.

The Passenger Pigeon

(*Above*). A photograph from an unknown source that is sometimes alleged to depict a living Martha. However, a careful comparison with the preserved specimen clearly reveals that this is, in fact, an early picture of her stuffed remains.

But their time together was not destined to last for long. Various zoos and institutions, alerted to the fact that these were the last surviving individuals, are reputed to have offered $1,000 (a considerable sum of money at the time) for them. Had an offer been accepted, it would have proved a bad bargain for the prospective purchaser. George died on July 10, 1910. It is not known if any part of his body was preserved, but probably nothing was. Reports state that his plumage was in a poor state!

From then on Martha was on her own, and she lived on in solitary confinement for four more years. It is easy to imagine that this situation might have been something of an ordeal for an individual belonging to a species as outrageously social as *Ectopistes migratorius*.

During this period of her life Martha got steadily slower and more infirm. At times she became so immobile that during busy periods her enclosure was roped off to prevent visitors from getting too close; otherwise they would throw sand and other objects to encourage her to move. Her cage measured eighteen by twenty feet, so there was little room for her to fly even if she had felt inclined to do so.

There are several reports of Martha during her last months. For his book *Hope Is the Thing with Feathers* (2000), Christopher Cokinos found an evocative description in a contemporary newspaper:

> *There will be no mistaking the bird, as its drooping wings, atremble with the palsy of extreme old age, and the white feathers in the tail make [her] a conspicuous object.*

Shortly before her death, an article in the *Cincinnati Enquirer* on Tuesday, August 18, 1914, told a sad tale:

> *The days of the last passenger pigeon…are now numbered…General Manager Sol A. Stephans…has abandoned hope of keeping it alive more than a few weeks longer at the very most. That it has been failing*

> *rapidly has been noted for some time, but it was not considered more than the feebleness of extreme old age until yesterday morning when Superintendent Stephans discovered it early in the morning lying on its back apparently dead. A few small grains of sand tossed upon it shocked it into activity again, and last night it was acting stronger and fed heartily when the evening feed was offered.*

Knowing that she was the last of her kind, Sol and her other keepers began to collect Martha's dropped feathers when she molted. They kept them in an old cigar box. When she did finally expire it was decided that her remains should be preserved. The *Cincinnati Enquirer* again took up the tale:

> *There will be no funeral for Martha. Instead, her remains, together with the feathers she has shed in molting, will be shipped to Washington, to be preserved in the Smithsonian Institution. Martha will…be shown to posterity, not as an old bird with most of her plumage gone, as she is now, but as the queenly young passenger pigeon that delighted thousands of bird and nature lovers at the Zoo.*

To this end, her body was frozen in a 300-pound block of ice to facilitate a successful passage from Cincinnati to Washington. Apparently, the journey took three days, at the end of which time the monstrous ice cube was almost entirely melted.

Skinning poor Martha presented something of a problem because she had been in mid-molt when she died. Also, pigeons are notoriously difficult

(*Facing page*). Martha today. At the time of writing, and after a gap of several years in the museum's vaults, she is back on display at the Smithsonian Institution. This photograph was taken by Donald E. Hurlbert and is reproduced courtesy of the Smithsonian Institution, Department of Vertebrate Zoology, Bird Division, National Museum of Natural History.

to skin even under the best of circumstances. Their feathers fall from their tracts rather easily because they are surprisingly loosely attached to the skin. This particular peculiarity is a device that sometimes enables survival. If an individual falls into the clutches of a cat, fox, or other predator, its feathers will drop with great ease. Should the predator have only a loose grip, this may enable the pigeon to slip from the clutches with the loss of just a few feathers. The trick doesn't always work, of course, and anyone who has chanced upon the remains of a wood pigeon captured by a wild animal will have noticed the mass of feathers scattered around.

(*Above, left*). William Palmer, who undertook the skinning of Martha's corpse.

(*Above, right*). Robert Wilson Shufeldt, who conducted research on her internal parts.

(*Facing page*). Nelson R. Wood, who stuffed Martha, with a pigeon and a Lammergeier.

So, taking no chances, the Smithsonian called upon a skilled specialist named William Palmer (1856 - 1921) to do the work. Once skinning was done, the internal parts were passed to Robert Wilson Shufeldt (1850 - 1934) for dissection and study, and the skin, with its remaining plumage attached, was given for stuffing into the care of Nelson R. Wood, the Smithsonian's head taxidermist.

And so the external appearance of Martha was preserved for posterity. For many years she remained on display but then was removed from public view, presumably as a sacrifice to scientific pretension. There seemed no awareness that this intrinsically interesting specimen would fascinate visitors and force them to confront the reality of extinction far more effectively than a thousand platitudinous lectures or pictorial displays on the subject. But times change and a new crop of museum workers have recognized the error of removing Martha from view, and at the time of writing she is back on display.

Art and Books

Art and Books

Of all the many existing images of the Passenger Pigeon, by far the most famous, and the one most regularly reproduced, is the picture created by John James Audubon. It was engraved as an aquatint and colored by hand, and used to illustrate Audubon's great masterpiece *The Birds of America* (1827 - 1838), now the most valuable of all illustrated books. The print was based on a watercolor that is owned by the New-York Historical Society. There is good reason for its popularity. It is quite simply a graphic image of the highest order, with an immediacy that strikes instant emotional notes in the viewer. Its visual grace and wonderful flow in terms of composition and subtlety of coloring are made all the more poignant by the fact that the species (unlike the image) no longer exists.

Despite these obvious virtues, a number of academic commentators have sought to downgrade it over a supposed lack of accuracy in the story it tells. Such naive criticism reveals one of the problems wildlife enthusiasts and scientists sometimes have when looking at works of art, and it underlines the reason fine art experts often take a more sophisticated view and sneer at a genre that has come to be known as "wildlife art." The great painter Edgar Degas (1834 - 1917) once wrote a few lines in French that translate as:

> *A painting is a thing that requires as much knavery, as much malice, and as much vice as the perpetration of a crime. Make it a lie and add the accent of truth.*

(*Previous two pages*). Detail from a mural in downtown Cincinnati, close to the site of the Cincinnati Zoo. Designed and begun by the well-known artist John Ruthven, it was completed in summer 2013 by several students (see also pages 146 - 147). Photograph by Maaike Goslinga and reproduced with her kind permission. © 2013 Maaike Goslinga | www.maaikegoslinga.nl.

(*Facing page*). Passenger Pigeon, female above, male below. Aquatint by John James Audubon and Robert Havell from Audubon's great work *The Birds of America*.

And this is exactly what Audubon has done. His positioning of the birds does not reveal how they would typically have behaved, rather the opposite. The picture is an artificial construct rather than the strict portrayal of truth so favored by most wildlife artists. It would have been normal for Passenger Pigeons to perch side by side on a branch rather than one above the other, and several commentators have taken Audubon to task over what they perceive to be an unforgivable fault. Such a pedantic analysis is well summed up by R. W. Shufeldt (1921), the man who dissected Martha and who, incidentally, held highly unpleasant and arrogant opinions on many aspects of life. Assuming that he knew more than Audubon about the aims and intentions of serious artists, he examined the picture and then wrote:

> *The error he committed in so many of his representations of birds is there repeated...the portraits of birds should never be shown in unusual poses or performing some action.*

Shufeldt obviously saw birds as entirely static entities! A less strident and more balanced analysis of the picture was made by Wallace Craig (1911), who acknowledged its fundamental beauty, and then made the entirely fair suggestion that the viewer should not be misled into believing it accurate in all its particulars. He pointed out that if food is passed from one bird to another, it is given by the male to the female and not vice versa, and, in any case, this was not necessarily behavior that *Ectopistes migratorius* engaged in. Nor, apparently, if absolute accuracy were to be achieved, should the tail of the male be spread. All of these are perfectly legitimate arguments. Nevertheless, Audubon's iconic image is the one against which all others can be measured, and it is the one that is indelibly stamped on the minds of those who have interested themselves in the species.

What the school of pedantic ornithology would have made of the awesome recent painting by Walton Ford (reproduced with his kind permission on pages 24 - 25) is anybody's guess. Nor would such critics have approved of Sara Angelucci's extraordinary images (reproduced opposite and on page 138).

(*Above*). A male Passenger Pigeon, from Sara Angelucci's series *Aviary* (2014). This image was produced by the artist in the manner of a nineteenth-century *carte de visite*. For information on the process used, go to: http://www.sara-angelucci.ca/aviary. © Sara Angelucci.

Despite his preeminence, Audubon was by no means the first to publish a depiction of the species. The earliest published image seems to be the one given in Mark Catesby's *The Natural History of Carolina, Florida and the Bahama Islands* (1731 - 1743). The original watercolor (see page 53) on which this colored engraving is based is owned by the British royal family, George III having bought it along with the rest of Catesby's watercolors in 1768. The book itself, like the watercolors and Audubon's masterpiece, is a collectible way beyond the means of any normal working person. Catesby had no artistic pretension and realized precisely just how limited his painterly skills were. Rather charmingly he wrote:

> *As I was not bred a Painter I hope some faults in Perspective and other Niceties maybe more readily [forgiven], for I humbly conceive Plants, and other Things done in a Flat, tho' exact manner, may serve the Purpose of Natural History better in some Measure than in a more bold and Painter like Way.*

This is, of course, the point that Shufeldt was trying to make, but expressed in an altogether more balanced and rounded manner, without any hint of criticism toward those possessing greater artistic ability. In another place Catesby wrote:

> *In designing the Plants, I always did them fresh and just Gather'd: And the Animals, particularly the Birds, I painted ...while alive (except a very few) and gave them their Gestures.*

The fact is that despite their obvious deficiencies, Catesby's images have retained their freshness and charm for almost three centuries.

A far more accomplished painter was a French woman named Pauline Courcelles (1781 - 1851), almost always referred to as "Madame Knip," a

---

(*Facing page*). A female Passenger Pigeon. A hand-colored engraving after a watercolour by Pauline Courcelles (Madame Knip) from her book *Les Pigeons*, published in Paris in parts between the years 1808 and 1811.

title assumed after marrying a landscape painter with that surname. Her artistic talents were considerable, and she became well known for her bird paintings, hand-colored representations of which decorate the pages of several extravagant tomes published in Paris during the early years of the nineteenth century. Among the best known of her books is *Les Pigeons* (1808 - 1811), a work that was a collaboration with the once-famous taxonomist Coenraad Temminck (1778 - 1858). A book with that title was sure to include the Passenger Pigeon, and painted pictures of a male and a female are featured. Clearly she was using stuffed birds as models, but the images produced have an intensity and power that arrest the viewer and, with their strange antiquarian quality, are quite unlike other pictures of the species. Her daughter, Henrietta Ronner-Knip (1821 - 1909), went on to become rather more famous than Pauline, and is well known today in art circles for her many paintings of domestic cats.

The first man to attempt to produce a full-scale work on American ornithology was a Scotsman named Alexander Wilson, who emigrated to the United States after controversy over a scurrilous poem he had written and published. This poem was a scathing attack on a local mill owner who was treating his employees harshly, and with the French Revolution in full swing, Scottish authorities were taking no chances. Afraid of the revolutionary fervor Wilson's attitude might inspire in their own land, local judges obliged him to burn copies of his poem in public and then gave him a short term of imprisonment. Naturally, perhaps, Wilson decided to make a fresh start in the New World. Leaving his home town of Paisley, he arrived in Pennsylvania and began work on the book he is famous for today. He started publishing it during the same year as Pauline Courcelles started work on her pigeon book, and it was finished just two years after she completed hers.

(*Facing page*). A male Passenger Pigeon. The second of Pauline Courcelles' (Madame Knip) hand-coloured engravings from her book *Les Pigeons*. The engraving was produced after one of her watercolors. Both of her images were clearly painted using stuffed birds as models, and it is doubtful that she ever saw a living individual.

Wilson gave his book the very appropriate title *American Ornithology*, and the first edition (1808 - 1814) stretched to nine volumes. Sadly, he died before completion of the eighth and ninth of these, and they were seen through the press by his friend the naturalist George Ord (1781 - 1866). In addition to a long and often-quoted account of the Passenger Pigeon, he produced an illustration. Lacking the flair and artistic ability of Audubon, his picture is, perhaps, a fairly ordinary piece of work, but it has obvious historical significance, and, notwithstanding Wilson's rudimentary drawing skills, it gives an unmistakably accurate impression of the appearance of the male of the species.

(*Above*). Male Passenger Pigeon with a Blue Mountain Warbler and a Hemlock Warbler. Hand-colored engraving by Alexander Wilson from his book *American Ornithology* (1808 - 1814).

Following these early depictions, many artists and illustrators of the nineteenth century chose to paint the species, and did so with varying degrees of success. Most of these efforts are characterized by their stiffness and the mediocre draftsmanship exhibited. Henry Meyer (1797 - 1865), for instance, accepting, perhaps wrongly, that the species had been seen in the British Isles, chose to illustrate it for his book *Illustrations of British Birds* (1835 - 1841). William Pope (1811 - 1902) produced a watercolor dated 1835 that is now in the Toronto Public Library and was used much later as the frontispiece for Margaret Mitchell's study *The Passenger Pigeon in Ontario* (1935). A Japanese artist by the name of Hayashi painted a picture that was highly thought of by the same Dr. Shufeldt who took Audubon so

(*Above, left*). Watercolor by William Pope (1835). Property of the Toronto Public Library.

(*Above, right*). Watercolor by a Japanese painter known only as K. Hayashi (circa 1900).

PASSENGER PIGEON.
(Male & Female)
Columba migratoria (Linn)
About two thirds of the Natural Size – Rare Visitant.

severely to task. It is a clear and attractive image, but little more can be said of it than that. Many similar examples could be cited.

At the end of the nineteenth century two artists blessed with much greater talents turned their attention to the species.

Louis Agassiz Fuertes (1874 - 1927), one of the most influential and prolific of American bird illustrators, produced several pictures, including a watercolor (see page 29) that he painted for William Mershon, who used it as the frontispiece for his monograph on the species, *The Passenger Pigeon* (1907). Another of Fuertes' paintings (see page 33) was clearly designed to show the different plumage states of male, female, and immature birds. Charles R. Knight (1874 - 1953), remembered today chiefly for his iconic images and murals of prehistoric creatures, also turned his hand to drawing a Passenger Pigeon on at least one occasion (see page 136).

(*Facing page*). Male Passenger Pigeon (above) and female (below). Hand-colored lithograph from H. Meyer's *Illustrations of British Birds*. Whether the species ever occurred naturally in the British Isles is debatable.

(*Above, left*). Louis Agassiz Fuertes. Photographer unknown.

(*Above, right*). Charles R. Knight. Courtesy of Richard Milner and Rhoda Knight.

(*Above*). A drawing by Charles R. Knight. Despite the fact that he is most famous for re-creations of prehistoric animals, Knight was well known for visiting zoos and drawing animals from life. This may be the only instance of a serious artist having drawn a Passenger Pigeon using a living bird as a model. The leaflet is dated 1903, and presumably Knight visited one of the three aviaries where individuals still survived shortly before this time.

Since the early twentieth century many wildlife artists have produced decorative, often romantic, interpretations of the species in all kinds of materials: paint, ceramics, wood, fabric. These are usually inspired by personal reaction to the Passenger Pigeon's story, and although often evocative and emotional in content, they usually add little to the historical record, nor do they possess the power of Walton Ford's image.

The direction of art moves on, of course, and contemporary artists like Ford adopt very different approaches to more traditional ways. A speciality of Greek artist Vaso Kafkoula, for instance, are images inspired by clockwork and metal, and one from this series features the species.

(*Above*). Passenger Pigeons by Vaso Kafkoula (2014). Vaso likes to draw things on scraps of paper, used postcards, and similar material. The label reads, "extinct because of humans." By kind permission of the artist and vasodelirium.blogspot.com.

Toronto-based artist Sara Angelucci's extraordinary images show Passenger Pigeon features morphed with those of humans. These are then bordered and framed so that the whole image looks like one of the calling cards known in the nineteenth century as a *carte de visite*.

Art and Books

There are many firsthand accounts of the Passenger Pigeon written by people who actually saw the birds in life, and these are scattered through antiquarian books of many types. Sometimes they are in the kind of works in which one would expect to find them: Audubon's *Ornithological Biography* or Alexander Wilson's *American Ornithology*, for instance. Other accounts appear in works that have no direct connection with the world of natural history.

Because of the intrinsic fascination of the story, selections from this wealth of material often appear in journals, newspapers, and magazines, and there are no end of articles in these kinds of publications briefly telling the Passenger Pigeon's tale and sometimes adding important information. Similarly, there are any number of books with a chapter or two featuring the species.

Passenger Pigeons are also featured in numerous poems, songs, and works of fiction, although usually these have little artistic merit. Perhaps the best-known work of fiction is a novel called *The Silent Sky* by Allan W. Eckert (1931 - 2011). First published during 1965, it traces the imaginary adventures (partly based on known facts) of several of the very last living individuals. Eckert specialized in this kind of material, having previously written a similar story called *The Last Great Auk* (1963), which weaves a tale around the lives of the last Great Auks (*Alca impennis* is the scientific name for the species, although recently it has become fashionable on less than satisfactory grounds to call it *Pinguinus impennis*). The vogue for this genre of novel had been inspired by a story written by a Canadian journalist named Fred Bosworth (1918 - 2012) and published in 1954. It was called *Last of the Curlews*, and instead of featuring Great Auks or Passenger Pigeons, it revolved around the lives of individuals belonging to another extinct North American species, the Eskimo Curlew (*Numenius borealis*).

(*Facing page*). The second of Sara Angelucci's fascinating images of Passenger Pigeons morphed into hybrid humans (for the male companion to this one see page 127). The picture opposite shows the female of the species and is taken from Sara's series *Aviary* (2014), which also features other extinct birds treated in a similar manner. The images are set within borders that give each the appearance of a nineteenth-century *carte de visite*. For information on the process used, go to: http://www.sara-angelucci.ca/aviary. © Sara Angelucci.

In addition to this kind of popular material, there have been several very thorough attempts to collect together all the known information that refers to the species, and assemble it into meaningful form.

The first of these attempts, *The Passenger Pigeon*, was published in 1907 and written by William Butts Mershon (1856 - 1943). Mershon had been a keen huntsman and fisherman, but at some point he realized that the pigeons remembered from his youth had virtually disappeared, and he began to use the fortune he had made from the lumber industry to fund various protectionist activities. He conceived the idea of producing a memorial to the species and eventually succeeded, writing modestly in the introduction:

> *For the last three years I have spent most of my leisure time in collecting as much material as possible...The result of this labor of love is scarcely more than a compilation, and I am under many obligations to those who have cheerfully assisted me...I wish that I might have been able to give them the more finished and literary setting that would have been within reach of a trained writer...I am merely a business man who is interested in the Passenger Pigeon because he loves the outdoors and its wild things.*

In many ways these words sum up the problems connected with any book written on the subject of an extinct bird. When stripped of pretension or supposition dressed up as fact, it can only really be a compilation of the work of those who have gone before.

The next substantial gathering together of information came in 1919 when the Altoona Tribune Company published a now-rare little treatise called *The Passenger Pigeon in Pennsylvania* by John C. French (1856 - 1934). The book is a strange amalgam of gossip, folklore, history, and personal recollection, with many sidebars on people from Pennsylvania who were somehow involved with the species.

During 1935 Margaret Mitchell (1901 - 1988) – not the better-known Margaret Mitchell who wrote *Gone with the Wind* – produced *The Passenger Pigeon in Ontario* under the auspices of the Royal Ontario Museum. As

indicated by the title, this work concentrates on records from Ontario, but nevertheless it is a remarkably thorough compendium of information, and it covers no fewer than 181 large pages.

(*Top, left*). John French working on his Passenger Pigeon book. Courtesy of Joyce Tice.

(*Top, right*). William Butts Mershon, author of the first book devoted solely to the species.

(*Above*). Arlie Schorger, author of perhaps the most comprehensive monograph.

(*Right*). The spine of John French's rare little book, published during 1919 in Altoona.

THE SILENT SKY: THE INCREDIBLE EXTINCTION OF THE PASSENGER PIGEON

ALLAN W. ECKERT
Author of THE GREAT AUK

HOPE IS THE THING WITH FEATHERS

A Personal Chronicle of Vanished Birds

CHRISTOPHER COKINOS

THE PASSENGER PIGEON

by **W·B·MERSHON**

Price, $5.00 Net

The Outing Publishing Company
NEW YORK

The Passenger Pigeon

A. W. Schorger

Exactly twenty years later, in 1955, Arlie Schorger (1884 - 1972) published perhaps the most comprehensive account in his book *The Passenger Pigeon*.

The volume resulted from many years of disciplined study and work. Long before the days of the Internet (which makes so much information so easy to access) Schorger toured libraries, art galleries, and museums to assemble everything that was known. He examined countless specimens and paintings. He interviewed anyone he could find who might have some new light to cast on matters. He collected together and combed through every scrap of printed material he could find, ranging from obscure newspaper clippings, all kinds of other ephemeral matter, leaflets, pamphlets, journals, and miscellaneous writings. And, of course, he studied many, many books. Some were common and easy enough to find, but at the other end of the scale, he often needed to consult excessively rare and obscure volumes that would take months, even years, to locate. Sometimes he had to persuade museums or libraries to allow him access to highly prized and expensive ornithological tomes that were in their care. Naturally, there were records that were available only in manuscript form, and it was necessary for Schorger to travel many miles and make many journeys in order to gather together all the information he needed to finish his self-appointed task. After its completion, his book quickly became the cornerstone for any future research into the subject.

Two other works from the middle decades of the twentieth century are worthy of mention. The first was published by the Wisconsin Society for Ornithology in 1947 and titled *Silent Wings: A Memorial to the Passenger Pigeon* (see pages 70 - 71). It features articles by several writers, including Arlie Schorger, and although produced in the form of a booklet it runs to forty pages, and is a thorough and valuable contribution to the literature.

An altogether more unusual publication is called *Where is that Vanished Bird?* This curious compilation was put together by Paul Hahn (1875 - 1962) and issued posthumously in 1963 by the same organization involved

(*Facing page*). The covers of four books that feature the species.

in the publication of Margaret Mitchell's work, the Royal Ontario Museum. It is simply an annotated list of all the preserved specimens of North America's extinct birds that Mr. Hahn was able to locate. He surveyed the world's museums (at least those willing to cooperate) and as many private collections as he could find, and then made copious notes. In other words, the book is a long, long list of stuffed birds, with a few records of skeletons thrown in. Mr. Hahn was a musician who became deeply interested in extinct birds and assembled many records connected with the subject. For the Passenger

(*Above*). The cover of Paul Hahn's extraordinary book.

(*Facing page*). Well-known painter of birds John A. Ruthven with the original picture that ArtWorks Cincinnati used as the design for a gigantic mural in downtown Cincinnati. See also pages 122 - 123 and 146 - 147. Courtesy of John A. Ruthven.

Pigeon he was able to locate no fewer than 1,533 stuffed specimens and cabinet skins. Doubtless there are a considerable number in private hands (and some provincial museums) that he didn't manage to hear about. He also managed to establish the existence of just fifteen skeletons, a surprisingly small number when compared with the number of stuffed examples.

In recent years two other very thorough volumes have been published. During 2000 Christopher Cokinos produced *Hope Is the Thing with Feathers*, a book that features not only the Passenger Pigeon but also other now-extinct North American birds. All of the chapters, including those that concern the Passenger Pigeon, contain a wealth of interesting and fascinating anecdotes and snippets of information, many of which throw new light upon the subjects covered.

Right at the start of 2014 Joel Greenberg published *A Feathered River Across the Sky* as part of Project Passenger Pigeon, a group formed to use the centenary of Martha's death as a focus to draw attention to the plight of endangered species today. Greenberg's richly detailed book is a useful addition to the definitive account by Arlie Schorger.

The Passenger Pigeon

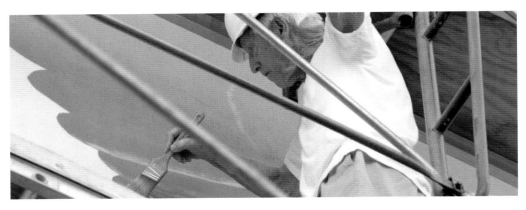

(*Above, top*). To commemorate the 100th anniversary of the death of Martha, work was begun on this gigantic mural in downtown Cincinnati. It was brought into being by ArtWorks Cincinnati during the summer of 2013, working in conjunction with John A. Ruthven, who kindly gave permission for use of the image in this book.

(*Above, bottom*). John Ruthven painting the enormous figure of Martha during one of his visits to the site.

(*Facing page*). The finished mural at Eighth and Vine in Cincinnati.

Art and Books

As might be expected, the 100th anniversary of Martha's death has stirred up a considerable amount of activity. In terms of art, perhaps the most fascinating is the mural that covers one side of a large building at the corner of Eighth and Vine Streets in downtown Cincinnati, close to the site of the zoo where Martha lived for years. Based on a picture by John A. Ruthven, the widely respected painter of birds who at the time of this writing is eighty-nine years old, the mural shows an imaginary Martha leading a flight of Passenger Pigeons as they wheel through the sky. Work began during the summer of 2013, and despite advancing years John mounted scaffolding to a height of forty-five feet to paint the figure of Martha. She measures twenty-one feet from beak to tail, and to complete this part of the work he climbed up every morning for two weeks, and on each visit painted for two or three hours. Even Michelangelo would have been impressed. When he had finished Martha (probably the largest image of a Passenger Pigeon ever painted), a group of students completed the work by copying the rest of his design in gigantic form.

The resulting mural is a spectacular and fitting tribute to the passing of one of North America's most charismatic and iconic species.

…Before sunset I reached L… …sburgh fifty-five miles. T… …in undiminished number… …three days in succession… …ns. The banks of the Oh… …d boys, incessantly shooti… …re flew lower as they pas… …re thus destroyed. For a…

Quotations

Quotations

The following quotations are from the writings of historical figures (many of them well known) who saw Passenger Pigeons in life.

Cotton Mather (1663–1728)

With more than 450 books and pamphlets to his name, Cotton Mather was one of the most influential of America's early religious leaders, and is probably the best known of all Puritan clergymen. His interests were varied, and Passenger Pigeons were among many diverse matters that caught his attention. Mather submitted several ornithological writings to the Royal Society of London (apparently in 1712 and 1716), but these did not receive full publication and remained for many years in manuscript form. He may have been the first to mention a milk-like substance that formed in the birds' crops.

(*Above*). Cotton Mather. A mezzotint by Peter Pelham, circa 1700.
(*Previous two pages*). Part of a page from J. J. Audubon's *Ornithological Biography*.

> *[I] take notice of vast Flights of Pigeons, coming and departing at certain Seasons: And as to this [I have] a particular Fancy of their repairing to some undiscovered Satellite, accompanying the Earth at a near distance…I will add a curiosity…yea one Man has…surprised no less than two hundred dozen in his barn, into which they have come for Food, and by shutting the door, he has had them all…The cocks were always by far ye fattest, and when we opened them we found in their craws, about ye Quantity of half a Gill of Substance like a tender Cheese-Curd. I asked Some of the Indians, what these Pigeons had eaten…They answered, It was nothing they had eaten, but something that came naturally into their crops, as milk does into the Dugs of other creatures.*

Pehr Kalm (1716–1779)

Pehr (often anglicized to Peter) Kalm was a Finnish-Swedish botanist and protégé of Carolus Linnaeus. He was sent to America in 1748 to collect plants new to science and kept extensive diaries detailing his travels.

(*Above*). The only known picture that may show Pehr Kalm. Painted during 1764 by J. G. Geitel, it has traditionally been regarded as a portrait of Kalm, but some modern authorities have doubted this.

In the spring of 1740, on the 11th, 12th, 15th, 16th, 17th, 18th, and 22nd of March...but more especially on the 11th, there came from the north an incredible multitude of these pigeons to Pennsylvania and New Jersey. Their number, while in flight, extended 3 or 4 English miles in length, and more than 1 such mile in breadth, and they flew so closely together that the sky and the sun were obscured by them, the daylight becoming sensibly diminished by their shadow.

The big as well as the little trees in the woods, sometimes covering a distance of 7 English miles, became so filled with them that hardly a twig or a branch could be seen which they did not cover; on the thicker branches they had piled themselves up on one another's backs, quite about a yard high. When they alighted on the trees their weight was so heavy that not only big limbs the size of a man's thigh were broken straight off, but less firmly rooted trees were broken down completely under the load.

The ground below the trees where they had spent the night was entirely covered with their dung, which lay in great heaps. As soon as they had devoured the acorns and other seeds which served them as food and which generally lasted for only a day, they moved away to another place.

A sea-captain...who had just arrived stated [he] had found localities out at sea where the water, to an extent of over 3…miles, was entirely covered by dead pigeons. It was conjectured that whether owing to a storm, mist, or snowfall [they] had been carried away to the sea, and then on account of the darkness…or from fatigue had alighted on the water…and met their fate

Extremely aged men stated that on three, five, or several more occasions in their lifetime they had seen such overwhelming multitudes in these places...so that 11, 12, or sometimes more years elapse between each such unusual visit.

Alexander Wilson (1766–1813)

Alexander Wilson was a Scottish poet, ornithologist, illustrator, and writer who came to America seeking fame and fortune. His multi-volume book *American Ornithology* (1808 - 1814) led to his becoming known as "the father of American ornithology."

> *The roosting places are always in the woods, and sometimes occupy a large extent of forest. When they have frequented one of these places for some time the appearance it exhibits is surprising. The ground is covered to the depth of several inches with their dung; all the tender grass and underwood destroyed; the surface strewed with large limbs of trees, broken down by the weight of the birds clustering one above another; and the trees themselves, for thousands of acres, killed as completely as if girdled with an ax. The marks of this desolation remain for many years…and numerous places could be pointed out, where, for several years after, scarcely a single vegetable made its appearance.*

(*Above*). A statue of Alexander Wilson in his hometown of Paisley, Scotland.

John James Audubon (1785–1851)

John James Audubon is quite simply the most important ornithological artist in history. His paintings may not be the most beautiful, but they revolutionized the art of painting birds. His massive double-elephant size folios are now the world's most valuable illustrated books. Unable to get his work published in America, he traveled to Britain to find subscribers for his book and engravers capable of turning his paintings into images suitable for publication. His account of the Passenger Pigeon is probably the most graphic of all written descriptions.

(*Above*). Detail from a painting of John James Audubon by G. P. A. Healy (1838). Oil on canvas. Museum of Science, Boston.

Whilst waiting for dinner at YOUNG's inn, at the confluence of Salt-River with the Ohio, I saw, at my leisure, immense legions still going by, with a front reaching far beyond the Ohio on the west, and the beech-wood forests directly on the east of me. Not a single bird alighted; for not a nut or acorn was that year to be seen in the neighbourhood. They consequently flew so high, that different trials to reach them with a capital rifle proved ineffectual; nor did the reports disturb them in the least. I cannot describe to you the extreme beauty of their aerial evolutions, when a Hawk chanced to press upon the rear of a flock. At once, like a torrent, and with a noise like thunder, they rushed into a compact mass, pressing upon each other towards the centre. In these almost solid masses, they darted forward in undulating and angular lines, descended and swept close over the earth with inconceivable velocity, mounted perpendicularly so as to resemble a vast column, and, when high, were seen wheeling and twisting within their continued lines, which then resembled the coils of a gigantic serpent.

Before sunset I reached Louisville, distant from Hardensburgh fifty-five miles. The Pigeons were still passing in undiminished numbers, and continued to do so for three days in succession. The people were all in arms. The banks of the Ohio were crowded with men and boys, incessantly shooting at the pilgrims, which there flew lower as they passed the river. Multitudes were thus destroyed. For a week or more, the population fed on no other flesh than that of Pigeons, and talked of nothing but Pigeons. The atmosphere, during this time, was strongly impregnated with the peculiar odour which emanates from the species.

It is extremely interesting to see flock after flock performing exactly the same evolutions which had been traced as it were in the air by a preceding flock. Thus, should a Hawk have charged on a group at a certain spot, the angles, curves, and undulations that have been described by the birds, in their efforts to escape...are undeviatingly followed by the next group that comes up.

About the middle of the day, after their repast is finished, they settle on the trees, to enjoy rest, and digest their food. On the ground they walk with ease, as well as on the branches, frequently jerking their beautiful tail, and moving the neck backwards and forwards in the most graceful manner. As the sun begins to sink beneath the horizon, they depart en masse for the roosting-place…

Let us now, kind reader, inspect their place of nightly rendezvous. One of these curious roosting-places, on the banks of the Green River in Kentucky, I repeatedly visited. It was, as is always the case, in a portion of the forest where the trees were of great magnitude, and where there was little underwood. I rode through it upwards of forty miles, and, crossing it in different parts, found its average breadth to be rather more than three miles…
The dung lay several inches deep, covering the whole extent of the roosting-place, like a bed of snow. Many trees two feet in diameter…were broken off at no great distance from the ground; and the branches of many of the largest and tallest had given way, as if the forest had been swept by a tornado. Every thing proved to me that the number of birds resorting to this part of the forest must be immense beyond conception. As the period of their arrival approached, their foes anxiously prepared to receive them. Some were furnished with iron-pots containing sulphur, others with torches of pine-knots, many with poles, and the rest with guns. The sun was lost to our view, yet not a Pigeon had arrived. Every thing was ready, and all eyes were gazing on the clear sky, which appeared in glimpses amidst the tall trees. Suddenly there burst forth a general cry of "Here they come!" The noise which they made, though yet distant, reminded me of a hard gale at sea, passing through the rigging of a close-reefed vessel. As the birds arrived and passed over me, I felt a current of air that surprised me. Thousands were soon knocked down by the pole-men. The birds continued to pour in. The fires were lighted, and a magnificent, as well as wonderful and almost terrifying, sight presented itself. The Pigeons, arriving by thousands, alighted everywhere, one above another, until solid masses as large as

hogsheads were formed on the branches all round. Here and there the perches gave way...with a crash, and, falling to the ground, destroyed hundreds of the birds beneath...It was a scene of uproar and confusion. I found it quite useless to speak, or even to shout to those persons who were nearest to me. Even the reports of the guns were seldom heard, and I was made aware of the firing only by seeing the shooters reloading.

Persons unacquainted with these birds might naturally conclude that such dreadful havock would soon put an end to the species. But I have satisfied myself, by long observation, that nothing but the gradual diminution of our forests can accomplish their decrease, as they not unfrequently quadruple their numbers yearly, and always at least double it.

In March 1830, I bought about 350 of these birds in the market of New York, at four cents a piece. Most of these I carried alive to England, and distributed amongst several noblemen, presenting some at the same time to the Zoological Society.

James Fenimore Cooper (1789–1851)

(*Above*). James Fenimore Cooper aged around sixty. After a photo taken by Mathew Brady.

James Fenimore Cooper is celebrated for his famous novel *The Last of the Mohicans* (1826), but he wrote many others that are rarely read today. One of these, *The Chainbearer*, contains a brief account of Passenger Pigeons. It is fictional but presumably based on a real experience.

> *The fluttering was incessant…Every tree was literally covered with nests, many having at least a thousand of these frail tenements on their branches, and shaded by the leaves. They often touched each other, a wonderful degree of order prevailing among the hundreds of thousands of families that were here assembled…Although the birds rose as we approached, and the woods…seemed fairly alive with pigeons, our presence produced no general commotion; every one of the feathered throng appearing to be so occupied with its own concerns, as to take little heed of the visit of a party of strangers, though of a race usually so formidable to their own.*

Chief Simon Pokagon (1830?–1899)

(*Above*). Chief Pokagon in a photograph used as the frontispiece for his book *Queen of the Woods*, published in 1899.

Simon Pokagon was the last chief of the Pokagon band of Potawatomi. He visited President Lincoln to insist on payment for the land on which Chicago was built, land that his father had sold. He was the author of a number of books and known as a great orator who spoke many times on behalf of his people.

> *The…pigeon…was known by our race as* O-me-me-wog. *Why the European race didn't accept that name was, no doubt, because the bird so much resembled the domesticated pigeons; they naturally called it a wild pigeon, as they called us wild men…When feeding, they always had guards on duty, to give alarm of danger. It is made by the watch-bird as it takes its flight, beating its wings together in quick succession, sounding like the rolling beat of a snare drum. Quick as thought each bird repeats the alarm with a thundering sound, as the flock struggles to rise, leading a stranger to think a young cyclone is being born.*

Mark Twain (1835–1910)

(*Above*). Mark Twain photographed during 1871 by Mathew Brady.

Samuel Langhorne Clemens, better known as Mark Twain, is one of the great American literary figures. Among his biographical writings he mentioned the Passenger Pigeon.

> *I remember the pigeon seasons, when the birds would come in millions and cover the trees and by their weight break down the branches. I remember the squirrel hunts, and prairie-chicken hunts, and wild-turkey hunts, and all that; and how we turned out, mornings, while it was still dark, to go on these expeditions, and how chilly and dismal it was, and how often I regretted that I was well enough to go...But presently the gray dawn stole over the world, the birds piped up, then the sun rose and poured light and comfort all around, everything was fresh and dewy and fragrant, and life was a boon again.*

Quotations

(*Above*). Plaque on a monument in Wisconsin. Photo courtesy of Kathleen Kaska.

Appendix

A Magnificent Flying Machine: The Anatomy of the Passenger Pigeon

by Julian Pender Hume

Much has been written about the abundance, decline, and extinction of the Passenger Pigeon but remarkably little about its anatomy. William MacGillivray (in Audubon 1831 - 1839: 34 - 35) described the digestive system and air passages, whereas Robert Shufeldt (1914) briefly described the osteology, and dissected Martha shortly after her death to examine the soft parts (Shufeldt 1915). Understandably, Shufeldt was quite sentimental about dissecting the last "Blue Pigeon," as he called the species, and after examining Martha's heart (which he declined to dissect) remarked:

> *With the final throb of that heart, still another bird became extinct for all time – the last representative of countless millions and unnumbered generations of its kind practically exterminated through man's agency.*

Shufeldt further noted, with some admiration, the great development of the Passenger Pigeon's pectoral girdle and wing bones.

Here the anatomy of the Passenger Pigeon is compared with that of the Rock Dove (*Columba livia*), a generalist pigeon, and an insular oceanic island species, the Pink Pigeon (*Nesoenas mayeri*) of Mauritius, in order to describe the differing anatomical characters exhibited in each of these pigeon taxa.

(*Facing page*). Two pencil drawings (Passenger Pigeon above and Pink Pigeon below) by Julian Pender Hume showing the great difference in wing shape and proportion between these species.

Musculature or Myology

The major flight muscles of a bird are the musculus pectoralis major and musculus supracoracoideus, which are primarily attached to the pectoral girdle. The smaller supracoracoideus is the muscle that elevates the wing during upstroke, whereas the much larger and more powerful pectoralis major is the primary muscle in downstroke. The latter can make up to 25 percent of a flying bird's weight. The great size of the flight muscles of the Passenger Pigeon was appreciated by Shufeldt when he dissected the body of Martha, and is indicative of a bird with a powerful flight.

Osteology

The pectoral girdle in a bird comprises the coracoid (shoulder bone), furcula or wishbone (fused clavicles), scapula (shoulder blade), and sternum (breastbone), providing attachment for the main pectoral flight muscles. This group of bones forms a rigid structure, countering the pressure that is created by the wing during flight. The coracoid and scapula loosely articulate with the head of the humerus (upper arm bone), which allows maneuverability of the humerus when flying. A canal (triosseal foramen) is created by this articulation, through which passes the tendon of the supracoracoideus, attached to the dorsal or top surface of the humerus and to the sternum. This acts as a pulley during the wing upstroke.

The coracoid is a paired bone that connects to the scapula and furcula and the anterior end of the sternum in a special slot (sulcus articularis coracoideus). In the Passenger Pigeon, the coracoid is especially large relative to the size of the bird, approximately the same length (33.4 mm) as that of the Rock Dove (34 mm) and Pink Pigeon (33.3 mm), but with the shaft straighter and the articulating ends much more robust.

The furcula consists of two clavicles fused at the distal end. It is situated in front of the sternum, connecting the coracoid and usually the scapula to the front of the sternum (apex carinae) via cartilaginous material. It acts as a strut and allows the scapula to move freely over the rib cage. In the Passenger Pigeon the furcula is sharply V-shaped and robust, with the articulating ends greatly expanded. In the Rock Dove it is similar but slightly less triangular and less robust, whereas it is elongate, gracile, and more U-shaped in the Pink Pigeon.

The blade-like scapula provides extra surface area for muscle attachment and lies across the dorsal surface of the rib cage. The scapula of the Passenger Pigeon is long, straight, robust, and enlarged at the distal end. The Rock Dove scapula is similar but shorter overall, but it is longer, more gracile, and less expanded in the Pink Pigeon.

The sternum is one of the largest bones of a bird, and its large surface area provides strong attachment for the flight muscles. It comprises a convex and concave basal plate with a ventrally directed keel or carina sterni, with the ribs connected to each side of the basal plate by small flexible ligaments. In fast-flying and migratory birds, there is a direct correlation between greater depth and forward projection of the keel and stronger flying capabilities (Düzler et al. 2006).

The Passenger Pigeon possessed an extremely large sternum, which can be better appreciated when comparing its body mass with other pigeons. The Passenger Pigeon was estimated to weigh between 255 g and 340 g, the Rock Dove weighs 238 to 380 g, and the Mauritian Pink Pigeon can weigh up to 410 g in the male (Blockstein 2002; Gibbs et al. 2001). The sternum of the Passenger Pigeon has an extremely deep keel (25 mm) supported by a robust, broad strut at the front (pila carinae), and has a distinct forward projection of the keel (apex carinae). The two spines (spina interna, above, and spina externa, below) on the front of the sternum, which articulate with the coracoids, are also long and robust. The sternum of the Rock Dove, a slightly heavier bird, is of equal size, but the depth of the keel (24 mm) and robustness of the pila carinae are somewhat less. The spinae are also less robust. In contrast, the sternum of the Pink Pigeon is overall extremely reduced, especially in the length and depth of the carina sterni (21 mm), and lacks the forward projection of the apex carinae. The spinae are also reduced, especially the spina externa.

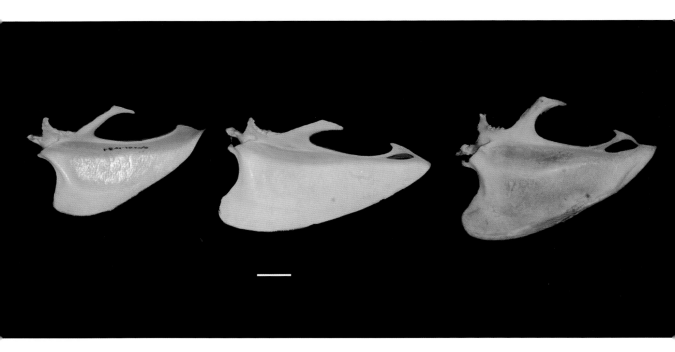

The ribs have small overlapping flanges (uncinate processes), which help stiffen the rib cage and prevent collapse during flight. As expected in a powerfully flighted bird, the uncinate processes in the Passenger Pigeon are extremely well developed.

The humerus in flying birds is generally short compared with the total length of the wing, as it has to withstand the pulling pressures placed on it during flight. The Rock Dove and Pink Pigeon humeri are overall much larger than that of the Passenger Pigeon, but in the lattermost it is comparatively more robust and straighter, with the groove for the tendon of the supracoracoideus more deeply excavated.

The radius and the ulna, provide support for the mid-wing, so it is no surprise that these skeletal elements in the Passenger Pigeon are extremely

(*Above*). A comparison of sterna of the Pink Pigeon (left), Rock Dove (center), and Passenger Pigeon (right). The comparative size of the sternum and the depth of the keel of the Passenger Pigeon are remarkable considering its slender build. The scale bar equals 10 mm.

well developed. The ulna is shorter than that of the Rock Dove and Pink Pigeon but straighter and comparatively extremely robust; the points where the secondary feathers attach (papillae) are also much more distinct. The radius is shorter in the Passenger Pigeon but comparatively more robust and straighter, especially when compared with the gracile radius of the Pink Pigeon.

The carpometacarpus or hand bone is fused for overall strength and support of the primaries, the main flight feathers,. The first digit or pollex supports the alula, a group of small feathers that control the air flow over the wing. As in the other wing bones, the carpometacarpus of the Passenger Pigeon is smaller but comparatively more robust.

Considering the size of its sternum, the bones of the wing in the Passenger Pigeon are somewhat surprisingly overall shorter than those of the Rock Dove and Pink Pigeon; however, their comparative robustness is a good indicator of a powerful flying ability.

The pelvic girdle and leg bones comprise the pelvis, femur, tibiotarsus, fibula, and tarsometatarsus. They combine to form a rigid structure to withstand the stresses of taking off and landing. The Passenger Pigeon regularly fed on the ground, and these elements do not differ, other than in smaller size, from those of the Rock Dove and Pink Pigeon, which are also both habitual ground feeders, the Pink Pigeon especially so.

Wing Shape

The shape of the wing can determine a bird's flying capabilities, and wings are generally divided into four types: elliptical wings, high aspect ratio wings, soaring wings, and high-speed wings. The shape of the wing can also reflect a bird's ecology. For example, an elliptical wing is short and rounded and ideal for maneuvering in dense vegetation such as in forests. This type of wing shape is exhibited in the Pink Pigeon. A high aspect ratio wing is much longer than it is wide, and can be seen in soaring birds, especially seabirds, and hovering birds such as kestrels, whereas soaring wings are used by thermal specialists such as vultures, pelicans, and storks.

The Passenger Pigeon has a typical high-speed wing, a feature shared with other fast-flying birds such as the Peregrine Falcon (*Falco pereginus*), in which the wings are longer and narrower than in a bird with an elliptical wing. The wings of the Passenger Pigeon are long and pointed (wing chord length to primary 220 mm; to secondary 120 mm), and it has a long-graduated tail, the two central feathers longer than the rest, whereas the trunk is comparatively narrow. This made the bird extremely streamlined. In complete contrast, the Pink Pigeon has comparatively short, rounded wings (wing chord length to primary 200 mm; to secondary 140 mm) and a broad tail, and the trunk of the bird is heavy and rounded. The Rock Dove has an equal wing length (wing chord length to primary 220 mm; to secondary 140 mm) to that of the Passenger Pigeon, but its wing is more rounded.

All of its adaptations combined to make the Passenger Pigeon one of the fastest-flying birds, and a species capable of covering thousands of miles when migrating, finding food resources, or when locating suitable nesting localities. Pigeons and doves have a faster rate of nestling growth than any other bird family and that the Passenger Pigeon was "born to fly" is reflected in the rapid development of its squab from hatching to fledging (see Blockstein 2002). The squabs were abandoned in the nests by the parents when thirteen to fifteen days old, but were able to fly efficiently within a few days after that, when they sometimes formed large flocks exclusively of immature birds (see Wilson 1808 - 1814).

The Passenger Pigeon, therefore, should not be remembered only as a tragic victim of human thoughtlessness and greed. In fact, it should be admired as a species that over millennia had evolved, to an exceptional degree, an anatomy conducive to a life on the wing; indeed, it was a magnificent flying machine.

Acknowledgments

Thanks must go to Julian Hume for writing the Appendix, and also for allowing his pictures to be used. Thanks should also go to my friends Raymond Ching, Walton Ford, Vaso Kafkoula, and Catherine Wallis for permitting the use of paintings and drawings. Maaike Goslinga kindly allowed use of her photo of the wall painting in downtown Cincinnati, and John Ruthven kindly allowed use of his images. Sara Angelucci, Maria Mangano, the late Nancy Tanner, the Denver Museum of Nature and Science, the Wisconsin Historical Society, the Smithsonian Institution, and the Lakeshore Museum Center also helped with images. Storrs Olson, Chris Cokinos, and Richard Milner provided assistance, and so too did Roddy Paine and his colleague Gavin Sawyer, who took several photographs for the book.

When an individual is seen gliding through the woods and close to the observer, it passes like a thought, and on trying to see it again, the eye searches in vain; the bird is gone.

John James Audubon

(*Above*). A Passenger Pigeon feather, actual size, drawn by Catherine Wallis (2013). Pencil on paper.

Further Reading

Audubon, J. J. 1827 - 1838. *The Birds of America*. London.

Audubon, J. J. 1831 - 1839. *Ornithological Biography*. London.

Blockstein, D. E. 2002. Passenger Pigeon (*Ectopistes migratorius*). In *The Birds of North America*, no. 611 (A. Poole and F. Gill, eds.). Philadelphia.

Boucher, P. 1664. *Histoire Veritable et Naturelle des Moeurs et Productions du Pays de la Nouvelle France, vulgairement dite le Canada*. Paris.

Carmichael Low, G. 1929. *List of Vertebrated Animals Exhibited in the Gardens of the Zoological Society of London, 1828 - 1927*. London.

Catesby, M. 1731 - 1743. *The Natural History of Carolina, Florida, and the Bahama Islands*. London.

Clark, E. B. 1901. *Birds of Lakeside and Prairie*. Chicago.

Cokinos, C. 2000. *Hope Is the Thing with Feathers*. New York.

Craig, W. 1911. The Expressions of Emotion in the Pigeons. III. The Passenger Pigeon. *Auk,* 28, 4: 408 - 427.

Craig, W. 1913. Recollections of the Passenger Pigeon in Captivity. *Bird-Lore*, 15: 93 - 99.

Deane, R. 1896. The Passenger Pigeon in Confinement. *Auk*, 13, 3: 234 .

Deane, R. 1908. The Passenger Pigeon in Confinement. *Auk*, 25, 2: 181-183, (April 1908).

Düzler, A., Özcan, Ö., and Dursun, N. 2006. Morphometric Analysis of the Sternum in Avian Species. *Turkish Journal of Veterinary and Animal Sciences* 30: 311 - 314.

Eckert, A. W. 1965. *The Silent Sky: The Incredible Extinction of the Passenger Pigeon*. Boston.

Fischer, M. 1913. A Vanished Race. *Bird-Lore*, 15, 2: 77 - 84 (March-April, 1913).

Fond du Lac Commonwealth. May 20, 1871. (Anonymous).

Forbush, E. H. 1913. The Last Passenger Pigeon. *Bird-Lore,* 15: 99 - 103.

French, J. C. 1919. *The Passenger Pigeon in Pennsylvania*. Altoona, Pennsylvania.

Fuller, E. 2000. *Extinct Birds*. London and Ithaca.

Fuller, E. 2014. *Lost Animals.* Princeton.
Gibbs, D., Barnes, E., and Cox, J. 2001. *Pigeons and Doves.* East Sussex (UK).
Greenberg, J. 2014. *A Feathered River Across the Sky.* New York.
Hahn, P. 1963. *Where is that Vanished Bird?* Toronto.
Hopkinson, E. 1926. *Records of Birds Bred in Captivity.* London.
Hume, J. P., and van Grouw, H. 2014. Colour Aberrations in Some Extinct and Endangered Birds. *Bulletin of the British Ornithologists' Club.*
Hume, J. P., and Walters, M. 2012. *Extinct Birds.* London.
Kalm, P. (1759) 1911. A Description of Wild Pigeons…in North America. Reproduced in *Auk,* 28: 53 - 66.
King, W. Ross. 1866. *The Sportsman and Naturalist in Canada.* London.
Kittredge, G. 1916. Cotton Mather's Communications to the Royal Society. *Proceedings of the American Antiquarian Society*, 26: 18 - 57.
Knip, P., and Temminck, C. 1808 - 1811. *Les Pigeons.* Paris.
Lincecum, G. 1874. The Nesting of Wild Pigeons. *American Sportsman*, 4: 194 - 5. (June 27).
Mactaggart, J. 1829. *Three Years in Canada: An Account of the Actual State of the Country in 1826, 1827 and 1828.* London.
Manlove, W. 1933. A Roost of the Wild Pigeon. *Migrant*, 4, 2 (June 1933).
Martin, E. T. 1879. Among the Pigeons. *Chicago Field* (January 25, 1879).
McIlhenny, E. A. 1943. Major Changes in the Birdlife of Southern Louisiana during Sixty Years. *Auk*, 60: 541 - 549.
Mershon, W. B. 1907. *The Passenger Pigeon.* New York.
Mitchell, M. H. 1935. The Passenger Pigeon in Ontario. *Contributions of the Royal Ontario Museum of Zoology*, no. 7.
Pokagon, S. 1899. *Queen of the Woods.* Hartford.
Révoil, B. 1856. *Chasses dans L'Amerique du Nord.* Tours.
Roney, H. B. 1879. Among the Pigeons. *Chicago Field* (January 11, 1879).
Schorger, A. 1955. *The Passenger Pigeon: Its Natural History and Extinction.* Madison, Wisconsin.
Scottow, J. 1696. *Massachusetts or The First Planter.* Boston.
Shufeldt, R. W. 1914. Osteology of the Passenger Pigeon. *Auk* 31, 3: 358 – 362.
Shufeldt, R. W. 1915. Anatomical and Other Notes on the Passenger Pigeon

Lately Living in the Cincinnati Zoological Gardens. *Auk* 32, 1: 29 – 42.

Shufeldt, R. W. 1921. Published Figures and Plates of the Extinct Passenger Pigeon. *Scientific Monthly*, 12, 5: 458 - 481 (May 1921).

Sullivan, J. 2004. *Hunting for Frogs on Elston, and Other Tales from Field and Street*. Chicago.

Temminck, C. (see Knip, P.).

"Tom Tramp". 1876. A Pigeon Roost. *Rod and Gun and the American Sportsman*, 8 (June 3, 1876).

Wilson, A. 1808 - 1814. *American Ornithology*. Philadelphia.

Wisconsin Society for Ornithology. 1947. *Silent Wings: A Memorial to the Passenger Pigeon*. Madison.

Wright, A. H. 1911. Early Records of the Passenger Pigeon. *Auk*, 28, 3: 366.

Wright, A. H. 1911. Other Early Records of the Passenger Pigeon. *Auk*, 28, 4: 427 - 434.

Wright, A. H. 1913. The Passenger Pigeon: Early Historical Records, 1534 - 1860. *Bird-Lore,* 15: 85 - 93.

Index

Abbott, J. 70, 72
Albersdoerfer, R. 17
American Museum of Natural History 99
American Ornithology 53, *132*, 139, 153
Angelucci, S. 126. *127, 138,* 138, 139
Atlantic Ocean 86
Audubon, J. *2-3,* 52, 54-5, 57, 92, 124-6, *125,* 128, 132, 135, 139, *154,* 154-7, 171
Auk, Great 12, 14-7, *17,* 139

Band-tailed Pigeon 32, *32*
Barnes, Mrs. C. 67
Bell Museum, Minnesota 66
Birds of America, The, *2-3,* 55, 57, 124, *125*
Boston 52
Bosworth, F. 139
Boucher, P. 52
Brady, M. *159*
Buttons 68, *68*

Canada 50, 79, 88
Carolina Parakeet *13,* 14
Cartier, J. 50
Catesby, M. 52, *53,* 128
Champlain, S. de 50
Chapman, F. 99
Chicago 9, 27, 44, 62, 65, 66, 92, 96, 97, 114
Chile 86
Ching, R. 86, *87*
Cincinnati 9, 27, 67, 92, 97, 109, 112, 114, 117-8, *122-3,* 124, 144, 146-7, *146, 147*

Clark, E. 66
Cockburn, J. 79
Cokinos, C. 34, 56, 64, 67, 68, 82, 117, *142*
Columbus 67
Connecticut 66
Cooper, J. Fenimore 157-8, *157*
Courcelles, P. 128, *129,130,* 131
Craig, W. 44, 46, 97, 126
Crested Bronzewing 98
Cross, L. 57
Cuba 37
Curlew, Eskimo *12,* 14, 139

Deane, R. 95
Degas, E. 124
Denver Museum of Nature and Science *84-5,* 86
Dinosaurs *16,* 17
Diplodocus *16,* 17
Dodo 9, 18, *18,* 19, 32
Dove, Mourning 32, *32,* 67, 83
Dove, Ring 97
Dove, Rock 162, 164-7, *166*
Duck, Labrador *12-3,* 14
Dudley, T. 51

Eaton, E. 31
Eckelberry, D. 34
Eckert, A. 139, *142*
Edwards, G. 19
Eldey, Island of 14
Elkington, Mr. 62
Eskimo Curlew *12,* 14, 139

Falcon, Peregrine 168
Fond du Lac Commonwealth 73
Ford, W. 23, *24-5,* 126, 137
Franklin County, Iowa 66
French, J. 140, *141*
Fuertes, L. *29*, 30, *33,* 135, 135
Fuller, E. 17, *17*

Geitel, J. *151*
George (Martha's companion) 114
George III 128
Goslinga, M. *122-3,* 124
Great Auk 12, 14-7, *17,* 139
Great Lakes, the 86, 88
Great Plains, the 88
Greenberg, J. 10, 68, 145
Grouw, H. van 39
Gulf of Mexico 86

Hadey Harbor, Labrador 14
Hahn, P. 72, 143, *144,* 144
Hardensburgh 155
Hartford, Kentucky 56
Hayashi, K. 133, *133*
Healey, G. *154*
Heath Hen *13,* 14
Herman, W. 112
Hubbard, J. 9, 43, 92, 99
Hume, J. *20-1,* 22, 39, *74,* 162-8
Hurlbert, D. *110-1,* 112, 118, *119*
Hurst's Stereoscopic Studies 81

Iceland 14
Illustrated Sporting News 75

Indiana 66, 68
Indian Ocean 18
Iowa 66
Iroquois 52
Ivory-billed Woodpecker 13, *14*, *15*, 34

Jones, O. 66

Kafkoula, V. 137, *137*
Kakapo 86, 87, *87*
Kalm, P. 151, *151*, 152
Kent 14
Kentucky 56, 158
Kentucky River 54
King, R. 56
Knight, C. 135, *135*, *136*
Knight, R. 135
Knip, Madame 128, *129*, *130*, 131
Kruse, M. T, and T. 68
Kuhn, J. 14, *15*

Louisville, Kentucky *155*

Mactaggart, J. 47
Mammoth, Woolly 9
Mangano, M. *61*
Manlove, W. 78
Martha 9, 67, 87, 97, 109, *110-1*, *112-9*, *113-6*, *119*, 121, 126, 145-7, *147*, 162, 164
Martin, E. *65*
Massachusetts 50, 66, 92, 97
Mather, C. 50, 150, *150*, 151
Mauritius 162
McIlhenny, E. 67
Melrose, Massachusetts 66
Mershon, W. 10, 30, 108, 109, 135, *140*, *141*, 142
Mexico 37
Meyer, E. 112, 115, *115*

Meyer, H. 133, *134*
Michelangelo 147
Michigan 57, 60, 64, 66, 82
Milner, R. 135
Milwaukee, Wisconsin 27, 92, 95, 96, 99, 114
Minnesota 66
Missouri 66
Mitchell, M. 133, 140-1, 144
Monroe County 82
Mourning Dove 32, *32*, 67, 83
Muskegon, Michigan 57

Naples 96
Natural History Museum, London 19, 39
Natural History of Carolina 52, 128
Nebraska 66
New England 88
New Jersey 66
New York 62, 88, 124
New Zealand 86, 87
North Carolina 66

Ohio 50, 57
Oklahoma 66
Ontario 66, 133, 140-1, 144
Ord, G. 132
Ornithological Biography 55, 139, *148-9*

Paine, R. *36*, 37, 57, *58-9*
Paisley 52, 131, 153
Palmer, W. *120,* 121
Parakeet, Carolina *13*, 14
Patagioenas 32, *32*
Pelham. P. *150*
Pennsylvania 22, 62, 66, 131
Peregrine Falcon 168

Petoskey, Michigan 62, *63*
Philadelphia, Pennsylvania 77
Pigeon, Band-tailed 32, *32*
Pigeon milk 42
Pigeon, Pink 162-7, *163*, *166*
Pike County 67
Pink Pigeon 162-7, *163*, *166*
Pokagon, S. 35, *158*, 158-9
Pope, W. 133, *133*
Popham, J. 5
Project Passenger Pigeon 145

Raleigh, W. 5
Revoil, B. 50, 56
Ring Dove 97
Rock Dove 162, 164-7, *166*,
Roney, H. 62, 64, 65
Ronner, H. 131
Roosevelt, T. 69
Rothschild, W. 39
Rundell, M. 82
Ruthven, J. 124, 144, *145*, *146*, 147

Salem 50
Salomonson, F. 72
Sargents, Pike County 67
Savery, R. 18, 19
Sawyer, E. 28, 29
Schorger, A. 10, 35, 56, 76, *141*, *142*, 143, 145
Scotland 52, 131, 153
Seattle, Chief 5
Seneca 73
Shawano Lake 95
Shelby, Michigan 60, 62
Shufeldt, R. *120*, 121, 126, 128, 135, 162, 164
Silent Wings *70-1*, 72, 143
Skuldt, H. *71*, 72
Smithsonian Institution 112, 118, 121
Sonny Boy 14, *15*

Southworth, P. 67, 68, *68*
Sparta, Wisconsin 38, 75
Stephans, S. and J. 112, 114, 117-8
Streptopelia 97
Sully, T. 54
Syme, J. 54

Takahe 86-7, *87*
Tanner, J. 14, *15*
Tanner, N. 14
Temminck, C. 131
Tennessee 66
Ten Sleep, Wyoming *16*, 17
Texas 66
Tokyo 96
"Tom Tramp" 60, 61
Toronto 133

Twain, M. 159, *159*, 160
Tyrannosaur 9

Utah 43

Virginia 69

Wallis, C. *5, 171*
Warren County 62
Washington, DC 118
Washington, G. and M. 114
Weidring, A. 99
White House, the 54
Whitman, C. 4, 9, 39, 43-4, 92, 95-9, *96*, 101, 106, 109, 114
Whittaker, D. 95-7, 99, 109, 114

Wilson, Alexander 52-6, *54*, 131-2, *132*, 139, 153, *153*
Wilson, Andrew 88, *89*
Wisconsin 66, *70-1*, 72-3, 75, 82, 143
Wisconsin Historical Society *1*, 10-1, *90-1, 93, 94, 95, 98*, 99, *100-7*
Wood, N. 121, *121*
Woodpecker, Ivory-billed *13*, 14, *15*, 34
Woods Hole, Massachusetts 9, 92, 97, 99
Woolly Mammoth 9
Wyalusing State Park *71*, 72
Wyoming *16*, 17

Zenaida macroura 32, *32*